Program Governance

Best Practices and Advances in Program Management Series

Series Editor
Ginger Levin

Program Governance

Muhammad Ehsan Khan

CRC Press
Taylor & Francis Group
Boca Raton   London   New York

CRC Press is an imprint of the
Taylor & Francis Group, an **informa** business

AN AUERBACH BOOK

CRC Press
Taylor & Francis Group
6000 Broken Sound Parkway NW, Suite 300
Boca Raton, FL 33487-2742

© 2015 by Taylor & Francis Group, LLC
CRC Press is an imprint of Taylor & Francis Group, an Informa business

No claim to original U.S. Government works

Printed on acid-free paper
Version Date: 20140625

International Standard Book Number-13: 978-1-4665-6890-7 (Hardback)

Library of Congress Cataloging-in-Publication Data

Khan, Muhammad Ehsan.
    Program governance / Muhammad Ehsan Khan.
        pages cm. -- (Best practices and advances in program management series)
    Includes bibliographical references and index.
    ISBN 978-1-4665-6890-7 (hardback)
    1. Corporate governance. 2. Project management. 3. Strategic planning. I. Title.

HD2741.K426 2015
658.4'04--dc23   2014022584

Visit the Taylor & Francis Web site at
http://www.taylorandfrancis.com

and the CRC Press Web site at
http://www.crcpress.com

This book is dedicated to my grandfather, Muhammad Yahya Khan, who was a professional educator for more than 50 years. Throughout his life, he guided and taught professionals who are now contributing to the development of nations. I pray for his departed soul!

Contents

SECTION II   Programs

Prologue

Organizations form programs to achieve strategic objectives that help them grow and perform more effectively and efficiently. The importance of these initiatives is such that an oversight function is required from executive management, whose role is to monitor the program ensuring that all parts are on track, support the program when required, and control different aspects of the program if things seem to go in the wrong direction. These dimensions of unification form the core of program governance.

The need for program governance comes from the fact that there are certain aspects of a program that are not under the direct control of the program manager, and he or she requires support from the program governors to ascertain successful delivery of the program objectives. In addition, the program governors have to ensure that the program activities are being carried out as planned, and that any major deviations are accounted for and approved. In the following chapters, for the sake of clarity, all governance entities or roles (such as steering committees, program boards, program sponsors, etc.) will be grouped together and referred to as program governors.

A few years ago, I was working with a team of software professionals to develop the first trading platform in the Gulf Cooperation Countries (GCC) region. The team, I still remember, could be called a *star team*. We were a start-up; we needed a team like that one. It was similar to a "band of brothers" working together to build their dream rather than a pool of professionals working toward an objective. We succeeded! We created our dream.

What happened after that is a memory to cherish. However, one of the key conclusions that we were able to get from the experience was the adaptability of the environment in which we worked. We called it *ordered chaos*. We knew we had to change as the context changed. What was important to us at one stage became insignificant at the later stage, so we changed ourselves. As the goals became clear, we shifted our approach. We continued to adapt until we reached our destiny. Since then I have been a strong supporter of using adaptable frameworks to manage teams and work. This book concentrates on presenting a program governance framework that is adaptable to the context of the program.

Section I focuses on establishing an understanding of governance. The chapters in this section discuss the concept of governance from different perspectives. This section will be useful for both practitioners and academics who want to understand the core of governance based on its underlying theories. While reading this section, you will see a number of references to research that has been carried out in the domain of governance. In addition, there is a detailed chapter on corporate governance, which has a strong influence on governance frameworks for projects and programs.

Section II defines programs and related concepts. A detailed discussion of temporary organizations is described in this section. The concept of the temporary organization is fundamental to understanding the concept of projects and programs. Such organizations are formed to meet certain organizational objectives for a limited period of time and cease to exist based on predefined conditions such as objective accomplishments.

Section III focuses on program governance and related ideas. This is where the complete framework of program governance is proposed from different dimensions. It proposes a new framework for the governance of programs based on changing context. This section also proposes guidelines for designing and implementing governance frameworks based on the proposed model.

Acknowledgments

Blessings from the Almighty Allah have been instrumental in my career and life.

I would like to thank my parents whose prayers and encouragement have always been a source of energy for me to progress in my life and to overcome obstacles that otherwise seemed impossible to surmount.

Many thanks and gratitude to my mentors and guides, Dr. Ginger Levin, Dr. Rodney Turner, Dr. Ralf Müller, and John Wyzalek, who have played a vital role in developing my authoring capabilities. Without their support and guidance I would still be a practitioner with ideas, yet unable to translate them into a framework.

I would also like to thank my brother-in-law Muhammad Jamal Maqsood, CISA, ITIL, for his consistent feedback. He read each and every chapter of this book and helped me to improve the overall quality of the content and context.

I would like to thank my grandparents, in-laws, brothers, sisters, elders, and friends, especially Dr. Abdul Waheed, Mohamed, and Saeed, who prayed for me, and who have consistently motivated and encouraged me during this journey.

Finally, a special thanks to my wife (Rafia) and children (Ayaan and Eshaal Maryam) for their patience, love, and encouragement. They are my source of motivation, and without their consistent support, the idea of writing the first book on program governance would still be a dream.

Author

Dr. Muhammad Ehsan Khan, Ph.D., is an entrepreneur and an internationally acknowledged professional on the subject of governance and management of strategic initiatives. An award-winning strategist with over a decade of leadership success, Dr. Khan is a founding member and presently serves as a partner and Vice President of Operations and Service Delivery for a UAE-based firm, Inseyab Consulting & Information Solutions LLC.

Dr. Khan did his Ph.D. in strategic, program, and project management (Major de Promotion/Valedictorian) at SKEMA Business School, France, and is a certified Program (PgMP) and Project Management Professional (PMP). He is also the recipient of the 2012 PMI James R. Snyder Award and was awarded the Young Researcher of the Year award by IPMA in 2013.

Dr. Khan is the designer/originator of Contingent Governance Framework for Projects (CGFP) and Contingent Governance Framework for Programs (CGFPrg). In his current capacity at Inseyab Consulting & Information Solutions, a business intelligence (BI) company, he has collaborated with BI experts to design a BI Framework for Project Portfolios (BIPPf). An initial version of this framework was presented and well received at the PMI Global Congress 2014, EMEA. In order to support other researchers and research publications, Dr. Khan works as a reviewer for the *International Journal of Project Management (IJPM)* and was awarded with a Certificate of Reviewer Excellence for his contributions.

With a special inclination toward strategic planning and governance of projects and programs, Dr. Khan has provided management, consulting, and mentoring services in the Middle East region. He has been involved in the establishment of PMOs, implementation of management/governance frameworks, and related practices and tools, in order to create an environment of project management excellence. He has also manages medium- to large-scale ICT programs and projects for various customers, especially in the government sector.

Section I

Foundation of Governance

1

Governance

"I want to penalize my vendors," was the response I got. I have been in the field of consulting for many years now. One of the questions that I ask my customers is to give me a reason for implementing a governance mechanism. While I was working with a large government entity, the department head responded that the vendors engaged in the project did not fulfill their contractual obligations, and he wanted to make sure that they were penalized accordingly. Even though I somewhat agreed with the frustration of the department head, implementing a governance framework for the sole reason of penalizing vendors did not make a strong business case. We need to understand the *what* and *why* of governance, so that we can articulate the reason for implementing a governance framework in a more objective manner.

The concept of governance has been used in organizations, as well as literature, from different perspectives. One perspective of governance is related to legitimacy and compliance of the undertaking. Corporations, and the entities governing them, want to ensure that organizational activities are legitimate and comply with rules and regulations, defined by the law, and ethical boundaries, defined by society. This perspective is also evident in the governance literature, which considers conformance to internal and external requirements as one of the most important functions of governance. This perspective is generally applied on an institutional level.

Another perspective, which is focused more on the actors than the actions, considers governance as a framework under which different stakeholders interact, within defined regulatory boundaries, in order to deliver an organizational strategy and objectives. Such frameworks do not manage the behavior of stakeholders; rather they lay the foundation for regulations and an environment in which the stakeholders can plan and execute their responsibilities.

To align our understanding of the term *governance*, this chapter will focus on defining governance, whereas Chapters 2 and 3 will explain the underlying theories of governance.

I recommend that readers go through these chapters, as the concepts discussed will be helpful in comprehending the models and frameworks discussed later in the book. There will be some thought-provoking material in these chapters that will help you understand the core aspects of governance.

DEFINING GOVERNANCE

Governance is considered an oversight function, which means that governance institutions, such as steering committees, should not manage the day-to-day operations of the entities they are governing but rather they should oversee and guide those activities. The creation of an organizing, decision-making, and control framework, where ordered rule and collective action are the foundation, turns out to be the basic objective of governance. Governance tools, such as contracts and agreements, define the scope within which an environment of self-regulation is created. Where there are multiple entities involved, governance mechanisms can help in identification of the roles and responsibilities of each, and more importantly, they can help clarify the power dependency between the entities.

Governance considers bounded rationality of human beings a major hurdle in successful delivery of objectives. Stoker (1998) takes into account the bounded rationality of humans and commented that "Governance means living with uncertainty and designing our institutions in a way that recognizes both the potential and the limitations of human knowledge and understanding" (p. 26).

The Office of Government Commerce (OGC) (2011) defines governance as a framework that defines the accountability and responsibility of people who are driving the organization, as well as the structure, policies, and procedures under which the organization is directed and controlled.

GOVERNANCE AT MULTIPLE LEVELS

While reviewing the work of Ralf Müller (2011), it was clear that the concept of governance was initially applied to steering countries, however, in order to streamline the process at all levels it was later applied to organizations. Governance can be defined on many levels such as international governance, public governance, corporate governance, and project governance.

International governance has to do with institutions set up to settle matters where several states are involved. In the case of international governance, the concept of self-governing networks can be applied where the states agree to behave under certain norms, rules, and regulations imposed by them on themselves. These regimes are formed by the network entities to resolve concerns of mutual interest and conflicts without resorting to undesired means. The issue of accountability arises when there is sharing of resources among participants of such regimes. This can, however, be resolved to some extent by the creation of some form of formal governance mechanism on top of the regimes, for example, the United Nations Organization or the World Bank.

Public governance is generally for initiatives and governance activities at the national level. Public governance is defined by the Organisation for Economic Co-operation and Development (OECD) (2005) as "Governance refers to the formal and informal arrangements that determine how public decisions are made and how public actions are carried out, from the perspective of maintaining a country's constitutional values in the face of changing problems, actors, and environments" (p. 16).

Governance is also carried out at an organizational level, which has to do with governance regimes set up in organizations, which ensure that the investment decisions are made correctly and that the decisions are executed properly and produce the required results. This is known as *corporate governance,* and O'Sullivan (2000) defines this concept as a "system of corporate governance shapes who makes investment decisions in corporations, what types of investments they make, and how returns from investments are distributed" (p. 1).

An organization generally meets its objectives outside the scope of normal operations through projects and programs. These initiatives are the basic means of implementing the strategy defined at the executive level. However, there are areas of management and control that are out of the

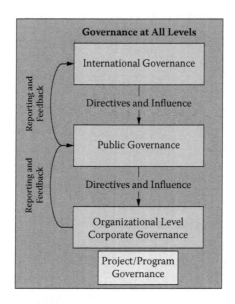

FIGURE 1.1
Governance at multiple levels.

domain of influence of project and program managers and come under the responsibility of executives. These responsibilities are executed through governance regimes. Program and project governance framework adheres to the corporate governance framework. However, there are certain other factors, such as attributes of the undertaking and external influences, that influence the program and project governance frameworks. These aspects will be discussed in Chapter 12.

As shown in Figure 1.1, each level of governance has an influence on the next level, whereas there is a reporting and feedback loop from a lower level to the level above it. There are certain directives provided by the preceding level that have to be considered while designing the governance framework at the level that follows. In addition, each level of governance provides input to the level above it by providing reports and feedback.

SUMMARY

Based on the definitions discussed in this chapter, governance is defined as a:

Framework under which actors perform their activities within a defined regulated boundary. The framework comprises legislation, regulations, policies, processes, and a value system that assists organizations to meet their goals. It, however, does not manage the actions and activities of actors; rather it provides them with the environment under which they can determine their strategy and actions in order to deliver benefits to stakeholders.

Before delving into the details of different governance mechanisms, it is imperative to recognize the underlying concepts and theories of governance, so that the practical as well as theoretical aspects related to governance can be described with greater clarity. The next two chapters will help us in achieving this objective.

REFERENCES

Müller, R. (2011). Project governance. In Pinto, J., Morris, P., and Söderlund, J. (Ed.), *Oxford Handbook of Project Management*. Oxford, UK: Oxford University Press.

Office of Government Commerce (OGC). (2011). *Governance*. Retrieved April 10, 2011, from http://www.ogc.gov.uk/delivery_lifecycle_governance.asp.

Organisation for Economic Co-operation and Development (OECD). (2005). *Modernising Government: The Way Forward*. Paris, France: OECD Publishing.

O'Sullivan, M. (2000). *Contests for Corporate Control: Corporate Governance and Economic Performance in the United States and Germany*. Oxford, UK: Oxford University Press.

Stoker, G. (1998). Governance as theory: Five propositions. *International Social Science Journal, 50*(155), 17–28. doi:10.1111/1468-2451.00106.

2

Transaction Cost Economics

My father was an officer in the Merchant Navy. Having spent 20 years sailing from one country to the other, he gained firsthand experience of different countries and cultures. He left his job because he wanted to ensure that his children had his undivided attention and started a small pharmaceutical distribution company. The company did well and became the largest distribution company, in terms of coverage in the province. At that time, I was 11 years old, but I was somehow attached to the business. I always asked my father two questions, as I was unable to comprehend:

1. Why the pharmaceutical companies did not distribute the products themselves? These companies had the financial resources and could develop the market more efficiently.
2. Why, my father, after making some money, did not get representation/agency of a pharmaceutical company? He had the market reach already, along with a solid distribution footprint. He could have brought a well-known product to the market and distributed it more efficiently.

I was given practical explanations from my mentors at different stages of my career; however, I really understood my father's rationale while reviewing transaction cost economics (TCE). The core reason was transaction costs.

The founder and chief developer of TCE theory, Oliver E. Williamson, received the 2009 Nobel Prize in Economic Sciences from King Carl Gustaf of Sweden, on December 10, 2009. TCE, which has its roots in the field of economics, focuses on proposing the most economical governance structure based on transaction attributes. This perspective suggests that the main purpose for nonstandard organizational structuring, that is, to buy rather than to build, is to economize on the transaction costs resulting from the transactions.

A transaction is said to occur whenever a product, good, or service is transferred between technologically separable interfaces (Williamson 1996). This means that the first entity completes its part of the work on the product, good, or service, and the next entity begins its work. In an ideal world, this transition should be seamless and without any problems.

Realistically speaking, entities face a number of issues, such as knowledge transfer, searching for information, issues of collaboration and understanding as well as commitment of different parties, while performing this transition. To reduce these glitches and perform transactions smoothly, certain transaction costs are incurred.

Transaction costs were first highlighted, with emphasis, in the classic paper, "The Nature of the Firm," when Coase addressed the question of emergence of firms. Coase (1937) defined *transaction cost* as the "cost of organizing" production (p. 390) and explained that this can include the cost of negotiation and finalization of contracts as well as the cost of price determination. It can also include costs of internal organizing and administration. These costs tend to be a determining factor, in which the organization decides whether the organization should carry out certain tasks by itself or exchange from the market.

Transaction cost can be deemed to be the economic equivalent of friction in physical systems (Williamson 1996). These are different from production costs, which are directly related to activities and materials for manufacturing, service, delivery, or product development. The total cost of a solution delivery or product delivery is a combination of production costs and the transaction costs. Production costs are considered to be the same in all economic systems. If we take an idealistic assumption of transaction cost absence of negligible transaction costs; then the organization of activities will no longer hold any value, as the advantage that one organizational mode holds over the other will be lost. Costless contracting and buying rather than producing will be the most cost-effective method of procurement. Also, if transaction cost did not exist, then the largest organization should be the most profitable organization since the coordination cost between functions would not exist. However, this does not happen because of the transaction costs, as companies have to deploy larger numbers of resources to organize the production competencies.

TCE weighs transactions from different perspectives, and the consequent hazards are identified. In addition, governance structures are studied from the dimensional lens, and different governance structures are identified, which assist in ex post hazard mitigation. The transactions which differ from each other based on certain attributes are aligned with the governance structures, which differ in their cost and structure resulting in the most economical result (Williamson 1998). These governance forms are based on contracts between parties and TCE deals with the identification and mitigation of the contractual hazards. The reason for focusing on this level of detail to economize costs is due to the understanding that "the analytical action resides in the details of transaction and governance" (Williamson 1996, p. 3). Thus, it is more rational to solve the problem, where it exists, instead of resolving the symptoms.

TRANSACTIONS—CORE ELEMENTS AND ATTRIBUTES

Transactions can be differentiated based on their attributes. TCE specifies three attributes, which classify transactions based on which governance structures are specified. These attributes are:

1. *Asset specificity*, which is related to relevancy of the asset to the task for which it is acquired or produced. It is also related to the redeployment of the same asset for other tasks, without loss of value or additional redeployment costs. Higher asset specificity will result in increased bilateral dependencies between technologically separable interfaces. TCE maintains that asset specificity can be of many forms and defines six forms, which are: site specificity, physical asset specificity, human asset specificity, dedicated assets, brand name, and time specificity.

2. *Frequency* of transactions, which defines the number of times the transaction occurs in general or is repeated between the parties. This frequency has a great impact on the relationship between the parties; for example, high frequency transactions between the parties over a long period of time can increase the bilateral dependency between them. Frequency of a transaction has a strong influence on the decision of whether some transaction should be performed within a firm or outsourced to the market.

3. The degree of *uncertainty* associated with the transaction, which means that one cannot predict with certainty how the future events

will unfold. This uncertainty can be due to bounded rationality, opportunistic behavior, and communication gaps or unclear environmental and related circumstances. Another source of uncertainty, which comes into play when two or more parties get into a bilateral (or multilateral) relationship and can be a result of opportunistic behavior, is behavioral uncertainty of parties, an example of which is strategically hidden or distorted information. Uncertainty can be a major reason for formation of a firm in the economy.

To have an economized governance structure aligned with the transaction, these attributes have to be taken into consideration. Ex ante incentive alignment as well as ex post governance structures should be aligned with these transaction attributes.

Behavioral Assumptions

TCE deviates from the orthodoxy by stating behavioral assumptions, which impact the transactions. These are *bounded rationality*, which states that humans do tend to make rational decisions. However, these decisions are bounded by their understanding, knowledge, and ability to process information, and *opportunism*, which is a calculated self-interest seeking behavior and may influence people to make use of unexpected circumstances in their favor (Williamson 1996).

While explaining the concept of firm, Coase (1937) refers to bounded rationality when he states "the difficulty of forecasting" (p. 392) while creating long-term contracts, which define the objectives and outputs but do not necessarily define the details of the work to be done to meet the objectives.

Complex contracts cannot be fully formulated because of this bounded rationality of humans. The potential of contractual hazards increase as a result of opportunistic behavior. Williamson (1996) states that "all complex problems of complex economic organization would vanish were it not for the twin conditions of bounded rationality and opportunism" (pp. 10–11).

Governance Structure Attributes

We have discussed above that the chosen governance structure, for a certain transaction, should have alignment between transaction attributes and the governance attributes. These governance attributes define the cost and competence of each governance structure. TCE specifies the following dimensions in which governance structures differ from each other (Williamson 1996):

1. Autonomous adaptation, which states that given the change in market conditions, such as price, the buyer and seller can adapt autonomously to maximize the opportunity or reduce the hazard.
2. Cooperative adaptation, where there is cooperation and planning needed between entities for adaptation to changing conditions.
3. Incentive intensity, which drives the buyer and seller to adapt according to the market conditions in order to gain the maximum advantage or motivates the employees within a hierarchy to move toward the objectives.
4. Administrative control, which defines the control applied by the organization in order to control the behavior of the team, has different applicabilities based on different governance models.
5. Each governance structure is supported by a different contract law regime.

Williamson (1996) classifies incentive intensity and administrative control as instruments or drivers that motivate parties to achieve objectives, whereas autonomous adaptation and cooperative adaptation as well as contract law are classified as performance attributes. These dimensions help in the classification of governance structures; which in turn assist in the implementation of governance structures aligned with transaction attributes.

HAZARDS OF CONCERN

The transaction attributes and behavioral assumptions defined above result in a number of ex ante and ex post hazards. The mitigation of these hazards is what TCE is concerned with and proposes through different means.

Incomplete Contracting

No matter how comprehensive a planning effort is invested in contract development, all contracts remain fundamentally incomplete. As discussed above, this contract incompleteness is due to intended rational behavior, which is bounded by various factors. Opportunism can also play a role in this case as the parties may strategically hide information that can assist in clarifying certain scenarios. This incomplete contracting poses a number of ex post strains between the contracting parties, which requires adaptation from them (Williamson 1996; Williamson 1998; Turner and Keegan 2001).

Bilateral Dependency and Fundamental Transformation

At the onset the buyer has a number of potential sellers competing to deliver the product or service. The seller also has no dependency on the buyer, as there is no relation present between them. However, once the buyer selects the seller, and both parties make transaction specific investments (asset specificity), the buyer and seller go through fundamental transformation and become bilaterally dependent. The seller now enjoys an advantage over the rivals, and both parties, which had no dependency on each other earlier now become bilaterally dependent on each other because of the contractual relationship. As mentioned earlier, contracts are fundamentally incomplete (as shown by the behavioral assumptions defined above); thus, the parties have to face the ex post contractual hazards, confront the problems at hand, and have to adapt cooperatively. In such a relation, identities of involved parties do matter, and they tend to work out solutions to the ex post hazards. Termination of the contract no longer remains a viable option, as the parties would have to incur substantial losses, due to the asset specific investments (Williamson 1996). This shows that bilateral dependency and fundamental transformation are positively associated with asset specificity.

Adaptation or Maladaptation

TCE considers adaptation to be the central problem of economic organization (Williamson 1996). Bounded rationality as well as opportunism can act as a major factor for maladaptation. Contracts align the incentives of the parties involved, however (as discussed above), because of bounded rationality, there can still be situations which need adaption from the parties involved. Thus, the main need is to craft governance structures that have superior adaptive properties. Markets enjoy the advantage of autonomous adaptation, where there is lack of control and dependency between parties, whereas cooperative adaptation is more applicable to hierarchies and hybrids (Williamson 1998).

There are other problems that we can discuss here, such as hazards of weak property rights or measurement hazards of multiple tasks; however, we can note that:

1. If we remove the behavioral assumptions of bounded rationality and opportunism, all of the hazards will vanish.
2. The transaction attributes have an effect on the existence as well as the magnitude of the hazards.
3. Even though contracts will be incomplete, however, hazards can be mitigated through ex ante incentive alignment and ex post governance.
4. Contracting and governance mechanism should be aligned with the transaction attributes.

LINKING GOVERNANCE, GOVERNANCE STRUCTURES, AND CONTRACTS

Moving from the concept of neoclassical firms as production function, in which more attention is paid to the inputs and outcomes and where organizations are treated as black boxes, TCE describes firms as a governance structure where the "allocation of activity between firms and markets is not taken as given but is something to derive" (Williamson 1996, p. 7).

Williamson (1996) defines governance as "the means by which *order* is accomplished in a relation in which potential *conflict* threatens to undo or upset opportunities to realize *mutual* gains" (p. 12).

Williamson (1996) treats institutions as a mechanism of governance, and states that TCE takes the institutional environment, which defines the rules of the game and operates at the organizational level, as given and works at the level of institution of governance, which deals at the level of transactions. The institutional environment creates the boundaries under which the institutions of governance have to be created and operated. Altering the institutional environment in order to economize is a complicated job; thus, better success can be achieved if the economizing is achieved through working a governance structure that is aligned with the transaction attributes. Governance structure refers to the institutional framework within which transactions take place.

Considering institutions to be a nexus of contracts (Jensen and Meckling 1976), transaction costs can be equated to the cost of contracting and governance will be mainly concerned with the ex ante and ex post contractual hazards and their mitigation (Williamson 1996). Müller (2011) mentions that TCE views organizations as a "network of contracts" (p. 302) where every contract defines the relationship attributes and associated governance structure.

Treating economization as the core problem of economic organization, TCE proposes that the contractual relationship and the governance structure should be aligned with the transaction attributes resulting in economization (Williamson 1996). A misaligned governance structure can create more problems, as it gives the parties a false belief of security, and once problems arise the structure does not provide the right environment for problem resolution.

Even though all contracts (especially complex contracts) are problematic (opportunism) and incomplete (bounded rationality), this should not act as an obstacle in defining the contractual relations. Rather, it should act as an opportunity for agents to look ahead, perceive threats, and account for them in the contracts. This can help them in mitigating contractual problems through ex ante incentive alignments as well as ex post governance mechanisms. Williamson (1996) calls this calculative-predictive-incomplete contracting "far sighted" (p. 9).

The concept that TCE advocates is that all alternative forms of contracting and governance are flawed. Thus, the comparison (when choosing the right form) should be made between these flawed feasible alternates instead of comparison with a hypothetical ideal. Thus, the form that has the highest net gains, as compared to the others, should be considered as the most efficient. Governance structures should be aligned with the transaction attributes and should provide the optimal balance of cost (economized) and control.

Another theme associated with governance is that of private ordering, which means that it is more cost-effective and efficient to have internal mechanisms (within the governance structure) to resolve ex post conflicts, which may arise because of the hazards defined in the previous sections. There is a general consensus that court ordering is a more mature and convenient option, however, court ordering should be used as a last resort. Parties that are part of the dispute know about the issues in more detail, and if given a private ordering mechanism within the governance structure, they would rather refer to that instead of moving to the courts for ruling. Courts do not deal with the internal organizational conflicts, thus a private ordering mechanism is needed. Private ordering should be in conjunction with the court ordering to form an efficient ordering mechanism, where the former must be invoked at first and the latter should be used as the last option.

THE IMPACT OF ASSET SPECIFICITY AND BEHAVIORAL ASSUMPTIONS ON CONTRACTS

The contracts are composed of four components, namely: the planning effort, which tends to take care of future consideration to meet the objectives; promise or commitment between the parties; pricing and related attributes, which is adjusted to competition; and finally governance mechanisms that govern the relations between the parties. TCE advocates that these contractual elements are in existence because of the behavioral assumptions and transaction attributes defined above. Absence of any one of the behavioral assumption or transaction attributes can result in the shift of focus of the contract from one element to the other.

Considering asset specificity, which is considered to be the most important transaction attribute, and behavioral assumptions as the determining factors, where uncertainty is assumed to be present in all conditions, Williamson (1985) explains the shift in contractual priorities and focus:

1. In the absence of bounded rationality, contracting becomes a planning effort, as everything can be perceived at the onset, and all hazards, including opportunism, can be dealt with and removed.
2. In the absence of opportunism, the focus shifts to promise and commitment, because even if there are contractual problems at a later stage that are due to the presence of bounded rationality or other reasons, the parties will not adapt to malpractices and will work together to achieve the required goals.
3. In the absence of asset specificity, the parties become independent of each other, as the buyer can buy from the market, and sellers face a more competitive market. Thus, the focus of contracting shifts to market competition.

4. In the presence of all the factors, the contract will remain incomplete (bounded rationality), the parties can show opportunistic behavior in case of an adaptation requirement, and the identity of the parties will matter (asset specificity), thus resulting in a shift of contracting focus on governance and related governance structures. An efficient governance mechanism may not eliminate all hazards; however, it can surely help in minimizing the impact of the hazards.

This shows the importance of governance in dealing with behavioral issues as well as transaction attributes. TCE is concerned with the creation of governance mechanism, which helps in economizing the organization, and where parties depend on private ordering and use courts as the last resort.

APPLICATION TO PROJECTS AND PROGRAMS

Accordingly to Williamson (1985), the application of TCE is broad, and any problem that can be formulated in terms of contract can be investigated from the lens of transaction cost economics. While reviewing the research, the validity of this statement was confirmed, as researchers tried to explain the behavior of buyers and sellers as well as the setup of their contractual relations in various industries using the TCE lens. To acknowledge the robustness of the application of TCE, the following discussion relates to the application of this perspective for temporary organizations.

Treating a project as a transaction, Turner and Keegan (2001) applied the TCE lens on projects, their uncertainty and the resulting association between the client and the vendor. This relationship is established between the client and the vendor because the client may not be willing to invest in the assets (assets required for project delivery such as human resources or tools required to produce certain deliverables) that will not be used again (nonspecific assets) and moves to the market to fulfill its needs. The project creates a state of bilateral dependency between the parties, as both parties have to be involved in the project throughout its life cycle. The uncertainty factor that exists in any project, as well as the bounded rationality of human beings, results in incomplete contracts between the parties. This, in turn, increases the bilateral dependency as both parties have to work together in order to resolve issues as and when they arise. However, there is also an opportunity for the parties to display self-optimizing opportunistic behavior in case of ex post adaptation needs. This situation necessitates a formal governance structure that should be in place for the whole duration of the project.

From the TCE point of view, the governance structure should not create an overhead, which does not justify its cost and should be economized based on the type of project. Projects have interlinked processes, and economization should be focused at the level of the project. This means that if we consider a project as a transaction and its processes subtransactions, than a subtransaction might not be economized intentionally (more resources are invested to reduce timeline), however, the result might create economy at the project level, which might not have been achieved otherwise. Turner and Keegan (2001) suggest that it is more economical to create a hybrid structure in case of projects. This is because projects need strong collaboration and communication between the buyer and the seller, thus a pure market-based structure cannot be applied where there is lack of bilateral dependency. At the same time, a project cannot exist with the administrative framework of the corporation as it has specialized needs, such as a strong collaboration between different functional entities and a high degree of uncertainty that requires quick adaptation. This requires a more collaborative hybrid structure instead of a hierarchical structure.

Müller and Andersson (2007) mention that the transaction cost perspective is generally applied at the individual project level in order to economize costs of the project; however, in the case of a portfolio of projects, it is more efficient to apply this perspective at the portfolio level to economize transaction costs at the portfolio level. They focused on the learning aspect within projects as well as the learning projects, where the focus is knowledge development, which can result in a reduction of transaction costs at the portfolio level as subsequent projects can have a broader knowledge about the customer, environment, and the requirements. They conducted open-ended interviews and participant observation on an enterprise resource planning project for a period of six months. Through this research they observed the impact of learning projects on subsequent individual projects as well as the portfolio. A risk that was evident in this case is that the client may decide to discontinue the contract in the immediate future, and the investment made by the supplier might incur losses. However, this risk can be mitigated by adding certain clauses in the contract through which the supplier's investment can be partially recovered or through mutual learning investments.

The research showed that investments made on learning projects may result in short-term losses such as cost and schedule overruns, however, in the long run, subsequent projects will reap the benefits through reductions in their transaction costs as they will not have to go through the same learning curve. If we consider projects to be subtransactions of the portfolio then transaction costs incurred at a certain subtransaction may result in economization of the parent transaction. Thus, it seems to be more appropriate to consider the learning costs (transaction costs) from a portfolio perspective, in order to economize on the overall portfolio, otherwise each individual transaction will have to incur this cost, in one way or the other, resulting in less than desirable economized results at the portfolio level.

SUMMARY

Institutions tend to systematize in such a way that they have structures and processes that can help them organize their work and resources in an efficient and cost-effective way. Because of the competitive environment that modern day organizations face, this focus has been on the rise. Transaction cost economics provides the framework, which can support organizations to achieve the desired efficiency in a realistic manner.

The advantage of the TCE perspective is that instead of focusing on the institutional environment, it focuses on the transaction level where economies of scale can be achieved within reasonable timelines and efforts. Thus, it provides a practical approach for organizations to economize based on the transactions, its attributes, and other associated factors. TCE also focuses on the fallibility of humans and takes into consideration certain behavioral assumptions, which increases its practicality in terms of implementation in the real world. That might be the reason why different researchers have tried to explain practical scenarios from the TCE perspective.

Organizations can apply the TCE approach to decide which transactions should be carried out that are internal to the organization, which transactions should form partnerships or consortiums, and which transactions should be outsourced to external parties. As the production cost will remain the same in any scenario, the decision in the end depends on which structure results in the most economized transaction costs.

TCE tends to move away from the notion of hypothetical ideals, rather, it focuses on flawed feasible options, which can provide an economized structure that is practical to establish.

One of the other salient attributes of the TCE approach is that it acknowledges that there will always be problems in contracting, and relations between parties will always be problematic. However, it also mentions that these risks can be mitigated through forming governance mechanisms and forward-looking contracts. The key concept of TCE is to economize on transaction costs by having governance structures, which differ in their cost and attributes, which align with transactions, and which differ in their attributes.

If we closely observe the work of Teo and Yu (2005), they added the notion of trust as a transaction attribute, which was not explicitly mentioned in the work of Coase (1937) and Williamson (1979). This shows that researchers are expanding on the foundations of TCE through the addition of concepts that are applicable to different scenarios and domains. For other researchers, TCE presents an opportunity to go beyond the governance structure and focus on the internal working of the governance mechanism and the dimensions of the governance such as surveillance, control, support, guidance, decision making, and strategic alignment. These dimensions can be studied by aligning them with governance structure and its attributes, such as the firm, which have more administrative control, as well as transaction attributes such as uncertainty, frequency, and asset specificity. The foundation has been established by Coase (1937) and Williamson (1979) on which researchers can build and provide explanations and solutions to different economization and related problems.

The diagram in Figure 2.1 combines the different concepts related to transaction cost economics thus providing a consolidated view.

This wide application of TCE validates Williamson's statement (1985) that the lens of transaction cost economics can be applied in any scenario, which can be explained in terms of contracts. He further mentions that TCE "is an empirical success story" (Williamson 1996, p. 20). However, one should be careful in overapplication of any theory or concept. That might be the reason that Williamson (1998) mentioned that the lens of TCE should be complimentary to other lenses, such as the lens of choice, rather than a substitute. This takes us to another, well-known, conceptual lens to look at governance, the agency theory that we will discuss in the next chapter.

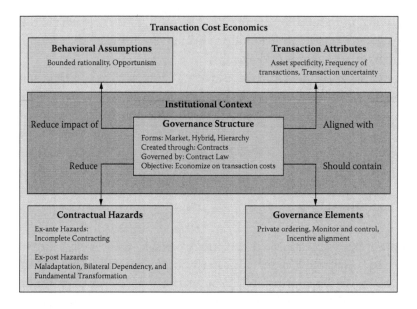

FIGURE 2.1

Transaction cost economics—conceptual model.

REFERENCES

Coase, R. H. (1937). The nature of the firm. *Economica*, New Series, 4(16), 386–405.

Jensen, M. C. and Meckling, W. H. (1976). Theory of the firm: Managerial behavior, agency costs and ownership structure. *Journal of Financial Economics*, 3(4), 305–399.

Müller, R. (2011). Project governance. In Pinto, J., Morris, P., and Söderlund, J. (Ed.), *Oxford Handbook of Project Management*. Oxford, UK: Oxford University Press.

Müller, R. and Andersson, A. (2007). Containing transaction costs in ERP implementation through identification of strategic learning projects. *Project Management Journal*, 38(2), 84–92.

Teo, T. S. H. and Yu, Y. (2005). Online buying behavior: A transaction cost economics perspective. *The International Journal of Management Science*, Omega 33(5), 451–465.

Turner, J. R. and Keegan, A. (2001). Mechanisms of governance in the project-based organization: Roles of the broker and steward. *European Management Journal*, 19(3), 254–267.

Williamson, O. E. (1979). Transaction cost economics: The governance of contractual relationships. *Journal of Law and Economics*, 22(2), 233–261.

Williamson, O. E. (1985). *The Economic Institutions of Capitalism*. NY: The Free Press.

Williamson, O. E. (1996). *The Mechanisms of Governance*. NY: Oxford University Press.

Williamson, O. E. (1998). The institutions of governance. *The American Economic Review*, 88(2), 75–79.

3

Agency Theory

When it comes to doing household work, such as changing a light bulb or fixing a running tap, I can be considered inefficient and unskillful. Please do not get me wrong. It is not that I do not want to do it. Or, perhaps I do not want to do it. Being unskillful, lazy, and uninterested in these activities forces me to hire skillful/interested labor to execute this work on my behalf. Sometimes when you get these people on board, your life can become "interesting." You have to:

1. Monitor their work from a quality/time perspective.
2. Ensure that they are not making a fool out of you by fixing problems that were never there.
3. Make sure that you are not being overcharged.

The moment you (the principal) hand over your responsibility/work to a contractor (the agent), you get into an agency relationship. Agency is defined as a relationship in which one entity (or person) appoints another entity (or person) to perform activities or services on its behalf. The foundational work on agency theory was performed by Jensen and Meckling (1976) when they mentioned the concept of principal and agent, where the former engages the latter to carry out tasks assigned on its behalf. This delegation of work is associated with delegation of decision-making authority and power to the agent. One of the basic assumptions of this theory is that humans are utility maximizers, which means that they will look after their own interests before taking care of other's benefits. Because of this divergence in goals and a self-interest-based mind-set, it is highly unlikely that the agent will take care of the interests of the principal at his or her own cost (Jensen and Meckling 1976; Donaldson and Davis 1991). Delegation provides the agent an opportunity to abuse this authority and power and work toward self-interests (Lupia 2001).

One of the major reasons for the formation of the agency is the principal's belief that the appointed agent has better capabilities and competence to carry out the task on his or her behalf, because of the agent's professional expertise, which will increase the value of the delegated asset or task.

These relationships are also created when the principal believes that the economic value of creating such a setup will be more than its current wealth function (Jensen and Meckling 1976). This can happen in the case where the principal wants to focus on some other domains and wants the agent to focus on the delegated area. Also in some cases, the work may have expanded so much that the principal cannot handle all the activities and thus delegates the work to an agent. Delegation of work allows the principal to perform more tasks that the principal can carry out independently.

Agency theory tries to address the agent's opportunist behavior and provides solutions to resolve conflicts of interest through creation and governance of effective contracts between the principal and agent (Eisenhardt 1989). Moldoveanu and Martin (2001) summarize this by stating, "Agency Theory is concerned with devising structural and behavioral measures that minimize inefficiencies in the contractual structure of the firm that arise from imperfect alignment of interests between principals and agents" (p. 4).

Agency theory involves actors (a principal and agent) and details on the relationship that exist between these actors when the agent acts on behalf of the principal. Because the agent has to perform activities for the principal, the agent is authorized by the principal to make decisions on the prinicipal's behalf in order to produce the desired results. This delegation creates a knowledge gap between the principal and the agent, as the principal cannot have the same knowledge that the principal would have if the principal were performing the tasks. This knowledge gap creates a sense of insecurity for the principal, as the principal is uncertain as to whether the agent is making effective decisions on the principal's behalf, and the outcomes are aligned with the principal's interests. To ascertain (to some extent) that the agent keeps the principal updated on the information and acts in the principal's interests, the principal creates certain incentives and monitoring mechanisms. Moldoveanu and Martin (2001) call them elements of the agency model:

1. Decision rights
2. Knowledge
3. Incentives—rewards and punishments

Ineffective combinations of the above causes the agency problems, which modern corporations are facing.

AGENCY PROBLEMS AND AGENCY LOSS

The basic agency problem is the conflict of interest between the principal and the agent as both may have entirely different goals when they get into the agency relationship. Another problem is the information imbalance between the parties resulting in problems of adverse selection and moral hazard (Moe 1995). Müller classifies these into ex ante, problems of adverse selection, and ex post, problems of moral hazard (Müller 2011). This information asymmetry can be based on the belief from the principal that he or she does not need in-depth information about the activities being conducted by the agent or based on the agent's understanding (or deception) that the principal does not need the information, or it will not be economical to provide information to the principal (Kortam et al. 2008). Following are the two main agency problems:

1. Ex ante adverse selection means that it cannot be said with surety that the agent, which the principal is selecting, will act in the principal's best interest or is capable of performing the task assigned. Adverse selection problems can be present in ex post conditions where the agent selects options which may maximize the agent's utility but are not in the best interest of the principal.
2. Ex post moral hazard problems occur when the agent acts in ways that may be more aligned to the agent's interest than the interests of the principal and produces results which are not aligned with the interests of the principal, and this information cannot be brought to the notice of the principal; thus the principal cannot take action. The moral hazard problem can be present in ex ante conditions as well, where the agent hides information from the principal that can help the principal to minimize the adverse selection problem.

Lupia (2001) states that the opportunity for the agent to behave as an opportunist is maximized when there is lack of alignment of interests, and the knowledge gap between the principal and agent is present. This is because the agent will be able to maximize his or her interests without having any fear of retribution, as the principal will have no information about the agent's behavior.

Eisenhardt (1989) also mentions the problem of risk sharing, which is because of the different risk attitudes. The subject of risk attitude is mentioned by Kortam et al. (2008) while reviewing business-to-business switching behavior. Because of this difference in risk, the attitude or the response that different parties may have toward a certain risk (positive or negative) can be entirely different, which in turn can cause decisions that are not aligned.

Karni (2004) believes that principals are concerned with outcomes as well as maintaning relationships with the agent through compensation, whereas the agent is concerned with outcomes to a limit that affects the agent's interests. Karni (2004) gives another perspective of this relationship when he states that the fatigue caused by the effort to produce the outcomes has a direct impact on the agent, whereas the principal only gets affected by it if it causes an alteration in the output. Thus, we can safely assume that the reaction of the agent and principal to this kind of fatigue can be entirely different and may motivate the agent to work toward his or her interests.

Moldoveanu and Martin (2001) mention two issues which cause the agency problem between the principal and agent:

1. Failures of managerial competencies, that is, managers failing to make an effective decision because of lack of expertise and experience.
2. Failure of managerial integrity, that is, managers failing to make an effective decision in order to increase their utility at the cost of the principal's utility function.

The problems identified above result in agency loss, which can be defined as the difference in the value of outcome and the best possible (best interest of the principal) outcome if the decisions were made by the principal. This loss can only be zero if the interest of the principal and the agent are fully aligned. Divergence in interests of the principal and agent increases the agency loss.

AGENCY COSTS AND CONTRACTS

The problems identified above and the resulting agency loss cannot be resolved or minimized unless the principal creates certain incentives for the agent and creates a mechanism in which the principal can monitor the activities of the agent. If the principal and agent share the same interests, and the principal is knowledgeable of the agent's actions and its results, agency loss can be minimized. However, it is close to impossible for the principal to ensure that the agent will make the best possible decisions on his or her behalf unless certain monitoring and bonding costs are incurred. Even then there will still be some conflicts between the decisions that the agent would make and those decisions that can be most beneficial for the principal. This can be due to utility maximizing behavior or bounded rationality of the agent. However, the interests of the parties can be aligned through contracts, which should be created in such a way that the agent will maximize his or her benefits if the agent acts in a way that is in the interests of the principal. These costs that have to be incurred to maintain a viable agency structure are known as the *agency cost*. These costs are comprised of:

1. Monitoring costs that are incurred to ensure that the agent behaves within a certain framework defined by the principal, and the principal observes the behavior and actions of the agent.
2. Incentive costs, which motivate the agent to align his or her interests with the principal.
3. Bonding costs by the agent to build trust with the principal. This creates a sense of security for the principal that his or her interests are safeguarded by the agent.
4. Residual loss, which is the loss incurred even though all possible measures have been taken by the parties.
5. The cost of creating the contract between the principal and agent to align their interests.

These costs are incurred to ensure that the agency loss (which is a by-product of agency relationships) is minimized to the best possible extent.

Contracts will assist in minimizing the agency loss, however, they are still deemed to be incomplete by nature. This may be due to the bounded rationality of human beings or because of the incomplete vision of future events or other factors beyond control. Another question that agency theory explains is the decision of choosing between behavior-based contracts or outcome-based contracts based on the situation at hand.

Eisenhardt (1989) proposes that if a behavioral contract is created between the principal and the agent and there are information systems available to the principal, with which the principal can monitor and verify the agent behavior, then there is a greater probability that the agent will tend to minimize opportunistic behavior.

The type of contract depends on a number of factors that can influence the behavior of both parties. These factors include, but are not limited to, clarity of objectives or outcomes, length of the principal–agent relationship, presence of measurement metrics, frequency of the principal-agent bonding, and riskiness of the outcome. This is in alignment with Eisenhardt's (1989) proposition in which she mentions that the type of contract (behavior-based or outcome-based) is based on certain situational attributes such as outcome uncertainty, risk aversion of the principal or the agent, goal conflict, task programmability, outcome measurability, and length of agency relationship.

It seems from the analysis above that the basic unit of analysis in agency theory is the contract, which exists between the principal and the agent when the agency relationship is being formed. Jensen and Meckling (1976) state that the modern corporation is a nexus of contracts. Eisenhardt (1989) also considers the contract to be the unit of analysis. According to Karni (2004), the principal–agent relationship exists under a contract. Moldoveanu and Martin (2001) believe that contracts assist in the resolution of agency problems through ex ante stipulation agreements.

However, Müller (2011) mentions that the unit of analysis in the literature as it has evolved is the collaboration between the principal and agent. Müller states that the initial literature focused on the contracts as the basic unit of analysis, however, researchers are now focused on social aspects of the relationships such as trust and mutual benefit. I agree with Müller's observation and believe that the contract is a tool for this collaboration rather than being the unit of analysis itself. The important aspect while designing a contract is the understanding that all parties entering into an agreement should look after each other's benefits in order to develop long-term relationships and create a win-win situation.

APPLICATION AND SUPPORT

Researchers coming from different backgrounds have applied the concept of agency theory in different setups and environments. This shows that the concept of the principal–agent relationship is universal such that it can exist between an employer and employee, head office and branch office, client and vendor, product manufacturer and distributor, and other arrangements. Eisenhardt (1989) states that "The heart of agency theory is the goal conflict inherent when individuals with differing preferences engage in cooperative behavior" (p. 63). To acknowledge the robustness of this theory, the following discussion reviews the application of agency theory in various scenarios.

While reviewing managerial behavior and ownership structure, Jensen and Meckling (1976) observed the behavior of the owner as the owner transfers the ownership to external parties. When the owner owns 100% equity, the owner makes decisions that maximize the owner's utility, both financial and nonfinancial, which in turn increases the organizational value. Now, when the owner sells some portion of the shares to an external party while still acting as an owner-manager, the owner-manager will continue to maximize his or her utility, but now the new owners will also bear the portion of the cost. As the owner-manager keeps selling shares to the new owners, the owner-manager's utility function with the organizational financial outcomes decreases, thus the owner-manager will try to maximize the overall utility by increasing nonfinancial benefits or perquisites.

There is also a possibility that as the owner-manager's shareholding value goes down, the owner-manager may avoid activities that require more effort on its part as compared to the benefit that the owner-manager is receiving from it. If the new owners want to monitor and control this behavior, they will have to expend some resources, which can minimize this behavior but cannot eliminate it. These monitoring costs, along with the incentives to the owner-manager, will create agency cost, which may reduce the market value of the firm (compared to the situation where there is a single owner managing the firm). However, these are necessary costs if the agency relationship has to exist. Both parties, that is, the owner-manager and the new owners entered into this relationship in the first place to increase their overall utility function (pecuniary and nonpecuniary).

Some of these costs have to be borne by the agent in order to increase the required utility of the parties as well as the organizational utility. In summary, a balanced ownership structure is needed where the utility of the parties involved should be fair if not maximized. This will ensure that the parties involved do not feel deprived or at a disadvantage, and will focus on maximizing benefits of the newly formed entity resulting in increased utility for all stakeholders.

Researchers have focused on the governance board and its formation from the perspective of agency theory. Donaldson and Davis (1991) state that having a principal, that is, governance board and agent, that is, chief executive officer (CEO) relationship in an organization, where the governance board monitors and controls the activities of the CEO, is not the most effective and efficient way to increase organizational utility and profitability. Based on a framework defined by agency theory, a governance structure will have a board of directors that will act as the principal and will curtail the opportunist behavior of the agent through different means. The principal will appoint a CEO as an agent to act efficiently on its behalf. Because of agency problems, agency costs have to be incurred; thus, the principal will create a monitoring structure to oversee the activities of the agent and will provide incentives to the CEO to act in their best interest. These incentives may include profit sharing or stock options in the organization, which may align the interests of the CEO (agent) with that of the board of directors (principal).

If the board includes a CEO as a chair, there is a strong possibility that the neutrality of the board will be compromised. Thus, based on the underlying principles of agency theory, the CEO and the board should be independent of each other. Having said that, the CEO should not be considered as a self-centered opportunist; rather he/she should have enough autonomy and authority to make effective decisions on behalf of his/her stakeholders. He/she should be part of the governance function as a chair who is at the center of decisions instead of someone who is always answerable to the board for the decisions he or she makes.

In sharp contrast to the above, Moldoveanu and Martin (2001) utilized agency theory in designing an effective board in order to make them "better agents for the stakeholders" (p. 2). The boards of directors act as a supportive layer between the shareholders and managers of the firm. They state that shareholders form boards of directors to enhance the "ratification, monitoring and sanctioning (reward and punishment)" (Moldoveanu and Martin 2001, p. 3), mechanisms that are required to curtail the opportunistic behavior of the agent and other agency problems. They state that inefficiencies in the modern corporation are due to the ineffective combination of agency model elements (defined above) along with the triple corporate veil (legal veil, information veil, and motivational veil) that exists at different levels of the organization and which hides the results of the decisions from the decision makers. To design an effective governance mechanism, it should be ensured that the decision-making authority is assigned at the level where the competence and knowledge are present, and people should be rewarded or punished based on the decision rights. Put simply, the authority, responsibility, accountability, and incentives for different roles should be aligned. There should be a monitoring mechanism present to observe the behavior in order to reward the incentives.

Moldoveanu and Martin (2001) believe that agency problems exist at the board level as there is a strong possibility that interests of the board may not be aligned to the stakeholders, and they may tend to increase their benefits on account of the stakeholders. In addition to this, the top manager may have an incentive to hide information from the board. Executives may also play a role in selection of the board members and may bring in people who do not have enough specific knowledge about the work domain or are "yes" people. To address these and other problems of the agency at the board level, Moldoveanu and Martin (2001) define mechanisms where they address all aspects of the board life cycle while designing effective boards of directors. This includes:

1. *Board selection mechanisms*: These address the question of how board members should be selected and appointed based on certain selection criterion and the priority of each one in comparison to the other. This addresses the problems of self-dealing on the part of existing board members and top managers (moral hazards resulting in adverse selection) who try to get people they desire on the board and imperfect presentations on the part of the prospective board member who tries to misrepresent the board about personal capabilities (moral hazards resulting in adverse selection).
2. *Board design mechanisms*: These deal with the processes and procedures that define the way decisions will be allocated to different roles so that the conflicts between board members and different layers within the organization are avoided, and the integrity of the decisions are not compromised (moral hazards and adverse selection).

3. *Board management mechanisms*: These answer the problems of board members acting in ways and making decisions that are not in the best interest of the stakeholders, such as delaying decisions, not voicing their concerns, dealing with uncomfortable situations, and other related issues (moral hazards which may result in adverse selection).

4. *Board compensation and monitoring mechanisms*: These deal with how the board members will be compensated based on their various factors. The compensation in the end should have the right mix of incentives (such as salary, cash bonuses, shares, and profit sharing), which will result in the reduction of ex post problems of board members working to increase their utility functions (moral hazards and adverse selection). In summary, there is a strong case that if the stakeholders take care of the interests of the board, the board will take care of the stakeholder's interests.

This comprehensive design covers most of the problems that may arise during the board's life cycle. We can consider an effective board design activity as a risk mitigation action, which may not resolve all agency problems but may avoid some problems altogether or mitigate a problem when it occurs.

Principal–agent relationships are present in government organizations as legislators delegate execution functions to departments and municipalities. Lupia (2001) suggested that the lawmakers, that is, parliament, delegates the actual implementation of policies and decisions to government agencies. The agencies employ the best resources in the area and thus can presumably execute the task more effectively. However, this creates the agency problem, such as a moral hazard and adverse selection (when the agencies abuse power) and related agency costs. Lupia (2001) states that if the agency problem realizes this, then there is a strong possibility of the lawmakers losing control resulting in anarchy. However, lawmakers can design a structure to reduce the agency problem such as having external audit bodies (whose interests are aligned with the principal) to be included in the process with the agents. Also, lawmakers in certain circumstances choose agents who share common interests such as bureaucrats who have an inclination toward their political party and its vision. This gives an insight into why governments appoint public servants based on political preferences. However, an ideal situation can be one in which the agents have the same political vision as well as interests and at the same time ensure they have the capacity, capabilities, and skills to carry out the required tasks.

Mudambi and Pedersen (2007) analyzed the shift of behaviors in multinational corporations (MNCs) from the application of agency theory to resource dependency as the organization evolves from central dependent units to international networks of functionally independent entities. Looking at it from an agency perspective they state that the managers at the headquarters act as principals, whereas managers at the subsidiaries act as their agents. The principal here delegates the decision-making rights at the subsidiary level to the agents who may have a tendency to pursue the interests of the subsidiaries they are representing as the interests of the subsidiary might not be fully aligned with headquarters. An example can be a decision from headquarters to pull out an important resource from a well-performing subsidiary and place the resource at another subsidiary. This decision might be beneficial for the MNC as a whole; however, it may not be in the best interest of the subsidiary. As the subsidiary has the decision rights loaned to them, they do not have the authority to disagree and have to comply with the decision. Because of situations like these, the subsidiary might have enough incentive to try to hide information from headquarters (moral hazard). Headquarters may create structures, provide incentives, and incur agency costs to reduce, control, and monitor such behavior. They state that the application of the agency perspective is applicable where the organization is:

1. Hierarchically structured.
2. Decision rights are loaned to subsidiaries.
3. Competence is managed and owned at the headquarters.
4. Subsidiaries are merely instruments of execution for headquarters' strategy.

As MNCs evolve to an organizational network of subsidiaries (instead of hierarchy), where the subsidiaries own the competencies and local resources, the application of resource dependency theory becomes more appropriate. In such conditions, the subsidiaries have bargaining power, which puts them in a position to make decisions on their own, which increases their utility function. These setups are considered to be loosely coupled where the headquarters may put up a framework of decision making but allow the subsidiaries to make decisions, which the subsidiary may deem necessary for its interests.

It seems that organizational leaders can select one of the theories based on its evolution; however, if one tries to apply it in real scenarios, the organization can apply both theories at the same time. One can apply agency theory in situations where organizations have certain subsidiaries under the hierarchy (owned by the headquarters), and resource dependency theory for others, which are part of the competencies creating subsidiary networks (such as locally owned franchises in different countries).

Turner and Müller (2004) focused on the information asymmetry aspect of the principal–agent relationship when they proposed the communication model between client (principal) and project manager (agent) in a project. They identified the communication needs of the client from the project manager and vice versa. They empirically proved that as the information asymmetry between the project owner and the project manager is reduced through communications, the problem of moral hazard and adverse selection reduces as well resulting in better project results. As communication between the parties involved increases:

1. The project manager (agent) starts understanding the communication requirements of the client in a better way. In addition, the project manager understands the vision and objectives related to the project more clearly, resulting in more effective principal–agent alignment.
2. The client (principal) becomes more comfortable with the knowledge of project decisions and results. The client also starts developing a sense of trust with respect to the project manager's skills and understanding about the project vision.

This improvement in project results may be because the project manager and the project owner start working in an environment based on trust and collaboration instead of one in which both parties have to incur agency costs to ensure the alignment of interests. We can conclude from their analysis that information asymmetry is one of the core issues that can in turn create other agency problems. Thus, agency problems can be minimized by having a work environment that is based on trust and strong collaboration.

CRITICISM

Researchers have also criticized this theory by stating that the basic assumption of human beings as utility maximizers cannot always be true, and financial gains are not always the only motivational factors.

Donaldson and Davis (1991), while reviewing the composition of the governance board and the involvement of the chief executive officer (CEO) as the chair of the board argue, based on empirical evidence, that even though there is widespread understanding of having the CEO separate from the governance board, it is more beneficial to unify the roles as it increases the return on equity and other gains to the owners. They empirically prove that the CEO duality as the executive manager and chair of the governance board provides greater returns to the stakeholders. They also show that this increase in return that is due to CEO role duality, not due to the long-term compensation of the CEO. This states that it is due to unification of the command chain and clarity of decision making that creates such positive results. Though the empirical evidence provided by them is not significant, it at least shows that combining the role is not as significant a problem as stated by other researchers and can actually be applied based on need. To maximize stakeholder returns they consider CEO duality (which is based on stewardship theory) to be a better option than separation of CEO and board chair role (which is based on agency theory). To summarize, we can state that instead of being concerned about the manager's self-centric behavior and controlling the behavior through different mechanisms, it would be more worthwhile to create an environment of empowerment and support to the manager (CEO) to generate better returns. They do not question the authenticity of the agency theory; instead they conclude that it cannot be applied in every situation, and other perspectives (such as stewardship theory) should be considered as alternatives based on the situation.

While applying the principal–agent relationship in projects, Turner and Müller (2004) state that the best results are achieved on projects when there is strong trust and collaboration between the client (principal) and project manager (agent) and that the project manager is empowered to make decisions. Strong collaboration and empowerment create an environment of trust where the client and project manager work together to achieve project success instead of looking after their own interests. The best performance on projects cannot be obtained through tight control, but instead by empowerment. The strong communication between the principal and the agent will eventually result in the minimization of information asymmetry and creation of a collaborative environment. This contradicts the agency theory perspective, which proposes that the way to align the agent's interests to the principal's is by creating a controlled environment and incentives for the agent resulting in agency costs (Jensen and Meckling 1976).

COMPLEMENTARY THEORIES

While reviewing the literature it was evident that there are complimentary theories, which give new insight into agency theory. While reviewing the work of organizational researchers, Eisenhardt (1989) concluded that agency theory is in alignment with a number of organizational theories and models. We will explore some theories that are either considered to be complementary to or a subset of agency theory.

Stewardship Theory

Stewardship theory considers that individuals, or in other words managers, act as stewards or custodians for the principals where they treat the interest of the firm and principals above their own individual interests and utility, and try to maximize the combined utility. Here the interests of the parties are aligned, and both are working toward the betterment of the organization itself. Müller (2011) mentions that the basic difference between principal–agent relationships and principal–steward relationships is that the former relationship is based on self-interest, and personal utility maximizing and deceitful behavior, whereas the latter is based on a foundation of trust, loyalty, and above all, individual gains. Müller's research considers stewardship theory as a subset of agency theory or at best, a complement to agency theory.

Donaldson and Davis (1991), while explaining the stewardship theory, question the basic assumption of humans being self-centric and financially driven. They base this theory on the understanding that managers do carry out tasks that might not be rewarding to them financially because of the sense of duty or loyalty to the organization. Managers or executives who are attached to the firm for long durations and have participated in the formation of the organization develop a sense of loyalty and attachment, which in turn creates an automatic alignment of goals and interests. Also, without any incentives being provided by the owners, the managers or employees of the organization develop an understanding that their future interests are dependent upon organizational success, thus their interests automatically align with the organizational interests. They state that the executive will be as effective as the organizational environment allows the executive to be. If the stakeholders provide a clear vision, have a clear expectation from the executive, and empower the executive to achieve the goals, the executive, or in other words the steward, will take care of their interests. From a comparison perspective, Donaldson and Davis treat stewardship theory as an alternative to agency theory not as a subset of it.

There seems to be a stark contradiction between agency theory and stewardship theory because of the underlying foundation difference of human behavior; however, if one looks deeper, the basic foundational behavior of utility maximizing is present in both theories. The difference is in the scope of utility maximizing; agency theory focuses on the utility maximizing behavior at an individual level by the agent, whereas stewardship theory states that the steward focuses on maximization of benefits for all stakeholders involved.

Having said all this, one cannot depend on any one theory. Principals cannot consider humans to always be opportunist and at the same time cannot blindly trust the person who is making decisions on their behalf. The combination of agency and stewardship should be balanced based on the circumstances faced by the organization or the situation at hand. A suggested model can be where the principal creates an environment of empowerment, clarity of roles and expectations as well as organizational vision, and conveys the same to the agent or steward. Principals can also create a supportive monitoring mechanism where the objective is to assist the executive in case the executive needs guidance in different circumstances instead of just focusing on punishing negative behavior. At the same time, the principal can take a supportive perspective and provide incentives to the executive not to align the interests of the executive but rather to reward non-self-centric behavior.

Resource Dependency Theory

Resource dependency theory is complementary to agency theory in the context of organizational structuring, decision making, and power confinement. According to this theory the unit, which owns the strategic resources and competencies required by the organization, will have the maximum power in the organizational setup (Mudambi and Pedersen 2007). Such units generally have an influence on organizational decisions, and at a minimum, they exercise a lot of bargaining power within the organization. Application of this theory is especially helpful in cases where decisions have to be made based on environmental conditions, and the resource competencies which are better recognized at the level of the unit such as subsidiaries operating in different countries. The negative consequence of such a setup is that the headquarters can become hostage to such power units. Thus, a balance must be created in order to gain advantages from such setups without risking anarchy in the organization. In MNCs, agency theory and resource dependency theory act as complementary concepts, and are applied as the situation demands.

SUMMARY

Principal–agent relationships exist in various settings, and agency theory provides a basis for an understanding of individual behaviors in such professional relationships. It shows how the negative impact of the agent's behavior can be curtailed or minimized so that the best possible outcome can be achieved in such relationships. From an organizational perspective, it provides insight into how organizational structures can be modeled in the hierarchy or in the market when dependent entities enter into a cooperative relationship, and how such structures can be governed in order to avoid problems of information asymmetry and goal alignment.

Agency theory treats humans to be self-serving individuals who show no consideration or interest for others. This point of view can be considered a bit extreme; however, the concept is not to act negatively toward individuals but rather to create structures and models that can produce the best results considering human fragility. The approach is to prevent negative behavior instead of regretting it at a later stage. Another negative aspect of agency theory is that it seems to be biased toward the principal and can promote exploitation of agents. However, as both parties know the negatives and positives of getting into such a relationship, it can assist both parties (principal and agent) to design contracts that can create a win-win situation.

The diagram in Figure 3.1 combines all the concepts related to agency theory thus providing a consolidated conceptual view.

The application of agency theory in various industries and research areas proves its universality. However, one theory cannot solve all organizational,

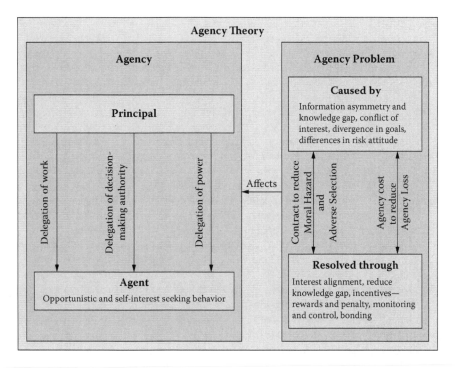

FIGURE 3.1
Agency theory—conceptual model.

managerial, and contractual issues; thus we need to consider the application of the theories based on situations and problems at hand. It is best to apply the agency concept in situations where there is a contractual agreement being created between different entities with diverging behavior, goals, and interests, and the objective is to maximize the outcomes.

Even though a general overview of governance, along with the detailed review of underlying theories of governance, gives us a broad understanding of the topic, the focus of this book is the governance of programs. This type of governance generally occurs within, or between, organizations. So it is important to understand governance considerations at the organizational level. The next chapter will focus on corporate governance, that is, the level of governance where programs are created.

REFERENCES

Donaldson, L. and Davis, J. H. (1991). Stewardship theory or agency theory: CEO governance and shareholder returns. *Australian Journal of Management, 16*(1), 49–65.

Eisenhardt, K. M. (1989). Agency theory: An assessment and review. *Academy of Management Review, 14*(1), 57–74.

Jensen, M. C. and Meckling, W. H. (1976). Theory of the firm: Managerial behavior, agency costs and ownership structure. *Journal of Financial Economics, 3*(4), 305–399.

Karni, E. (2004). *Axiomatic Foundations of Agency Theory*. Academic Paper, Baltimore: Johns Hopkins University.

Kortam, W. A., Aish, E. M. A., and Hassan, S. S. (2008). *Using Agency Theory in Understanding of Switching Behaviour in B2b Service Industries "I."* Working Paper Series. Working Paper No. 6. Faculty of Management Technology. Cairo: German University in Cairo.

Lupia, A. (2001). Delegation of power: Agency theory. In Baltes, P. B. and Smelser, N. J. (Ed.), *International Encyclopedia of the Social and Behavioral Sciences* (Vol. 5, pp. 3375–3377). Oxford, UK: Elsevier Science Limited.

Moe, T. M. (1995). The politics of structural choice: Toward a theory of public bureaucracy. In Williamson O. E. (Ed.), *Organization Theory from Chester Barnard to the Present and Beyond* (Expanded Edition, pp. 116–153). New York: Oxford University Press.

Moldoveanu, M. and Martin, R. (2001). *Agency Theory and the Design of Efficient Governance Mechanisms*. Working Paper, Rotman School of Management. University of Toronto.

Mudambi, R. and Pedersen, T. (2007). Agency theory and resource dependency theory: Complementary explanations for subsidiary power in multinational corporations. In Pedersen, T. and Volberda, H. (Ed.), *In Bridging IB Theories, Constructs, and Methods across Cultures and Social Sciences* (pp. 1–16). Basingstoke: Palgave-Macmillan.

Müller, R. (2011). Project governance. In Pinto, J., Morris, P., and Söderlund, J. (Ed.), *Oxford Handbook of Project Management*. Oxford, UK: Oxford University Press.

Turner, J. R. and Müller, R. (2004). Communication and cooperation on projects between the project owner as principal and the project manager as agent. *European Journal of Management, 22*(3), 327–336.

4

Corporate Governance

WHY CORPORATE GOVERNANCE?

I have always been an avid lover of cricket, watching and playing cricket for years now. Twenty-five years ago, I bought a cricket kit. It was the Eid holiday (an important Islamic holiday celebrated by Muslims throughout the world). I had been able to gather enough money from my elders on Eid to buy a kit for myself. When my father came to know about this, he was infuriated. At that time I was not sure why he was getting angry. It was not the money aspect; he used to spend a lot for our family. It was not the cricket kit; he was a cricket lover himself.

I realize now that I acted in a manner that was noncompliant with the governance framework that ran in our family. There were certain items that we could buy without asking permission from our father. However, buying a cricket kit was a decision where he needed to intervene and grant permission. If I had consulted with him, he would have taken me to the shop and gotten one for me. Too late, money spent and wrath bought.

It is understandable that every institution, whether family or corporation, needs some sort of a framework under which people can operate and make decisions. This is the reason why corporate governance has been around for years; however, because of a number of corporate failures during the 2000s (e.g., Enron in North America and Parmalat in Europe), it has become a discipline in its own right.

Corporations realized that even though sustainable financial results are the core of progressive organizations, they could not be achieved with a singular focus on financial aspects. People also began to realize that the existing controls in an organization, which were thought to be sufficient to safeguard the interests of shareholders, were not adequate—hence some reforms were required.

Corporate governance has a significant impact on the economic health of corporations and society. Corporate fraud, scandal, and malpractice have renewed interest in corporate governance.

Principal–agent relationships, or in other words, the separation of ownership and control, seem to be one of the fundamental reasons for having a corporate governance framework in place. The purpose is to ensure that the interests of the owner or shareholder (principal) are safeguarded by the executive management (agent). The Organization for Economic Cooperation and Development (OECD) principles of corporate governance focus on the problems that arise because of separation of ownership and control. These principles also refer to the protection of shareholder interests as a major aspect of corporate governance.

The principal–agent relationship cannot be managed by introducing an excessive monitoring mechanism, but a combined mechanism of incentive-based contracts and a monitoring system may be needed, which requires the forming of a corporate governance framework.

The focus of corporate governance should not only be to manage an organization's current state, but it should also focus on taking the organization to a better future state. Governance board members act as stewards to the principal as they have the role of protecting the interests of the stakeholders in the corporation.

DEFINING CORPORATE GOVERNANCE

The initial focus of corporate governance was on shareholder interests; however, increasing shareholder value may result in affecting the wider range of stakeholders who will be affected directly or indirectly. Organizations have an ethical obligation to consider the society they are operating in, and hence they have social obligations toward a wider range of stakeholders. This results in greater focus toward "stakeholder theory" instead of "shareholder theory." OECD, while setting up the principles of corporate governance, also focuses on a wider range of stakeholders. Thus, stakeholder interests are at the core of corporate governance. Corporate governance is defined by the OECD (2004):

> Corporate governance is one key element in improving economic efficiency and growth as well as enhancing investor confidence. Corporate governance

involves a set of relationships between a company's management, its board, its shareholders and other stakeholders. Corporate governance also provides the structure through which the objectives of the company are set, and the means of attaining those objectives and monitoring performance are determined. Good corporate governance should provide proper incentives for the board and management to pursue objectives that are in the interests of the company and its shareholders and should facilitate effective monitoring. (p. 11)

This definition shows that corporate governance is not just focused on the interests of the organization and its shareholders; rather, that it considers the relationship with all stakeholders as well as their interests. It also shows that corporate governance at the organizational level results in improving economic conditions of the market. Corporate governance is a system that defines how the organization should be directed and controlled. It has to be understood that corporate governance is not simply an internal-looking regulatory function; rather that it involves consideration for external stakeholders, such as the market, as well as the industry standards.

From a principal–agent relationship perspective, corporate governance focuses on streamlining the relationship between the principal and the agent through monitoring mechanisms, such as transparency, compliance, and reporting as well as board and shareholder composition.

Corporate governance is also about allowing managers to drive the company forward; however, this freedom is under a framework of control mechanisms and accountability, decision-making process, and clear distribution of power (remember my cricket kit!).

Corporate governance at the organization level is thus concerned with:

1. Setting up policies, controls, and procedures for provisioning and the management of organizational assets (including employees) and services in order to maximize organizational benefits.
2. Creating a decision-making mechanism and assigning decision rights to the correct level.
3. Establishing practices to meet internal and external requirements related to effectiveness, efficiency, confidentiality, integrity, availability, compliance, and reliability for its information and information-based services.

In addition, the legitimate rights of the shareholders, in particular, and the stakeholders, in general, should be clearly defined within the governance framework. Chau (2011) summarizes corporate governance framework by mentioning that "The heart of all instruments and mechanisms should be directed to proper stewardship, integrity, openness, transparency and accountability without excessive surveillance and bureaucracy" (p. 10).

Within the conformance, performance, and relating responsibility (CPR) framework, Pultorak (2005) focuses on the board ownership of corporate governance and mentions that "corporate governance must be ordered within a framework established by the board that aligns and informs day-to-day decision-making, objective setting, achievement monitoring and, communication" (p. 292).

Corporate governance and organizational well-being is the responsibility of the board. Good governance is not just about compliance and transparency; rather, it benefits organizations by increasing the confidence level of the market on the organization, which eventually results in higher profitability. Thus, corporate governance should be seen as a strategic tool instead of an audit mechanism.

FACTORS AFFECTING CORPORATE GOVERNANCE AND ITS EVOLUTION

Only one thing is observed to be constant in our world: change. It implies that the requirements of governance have to evolve to meet stakeholders' needs. The core of the governance perspective is its capacity to accommodate to a changing environment and to use processes of governance within an organization.

Various environmental, external, and internal factors have contributed toward the continuous evolution of the concepts that form the foundation of corporate governance. Organizational complexity is increasing as most contemporary organizations regularly interact with multiple interfaces, such as vendors, partners, and other legal entities, where each entity might be working toward increasing value for the organization. This rise in complexity and interdependence results in a greater need for corporate governance. The relationship between entities also creates the question: Who should perform governance?

Historically, corporate governance's focus has been on financial measures and management. Even though the financial bottom line still remains the main business objective, factors other than financial ones may have an impact on an organization. Thus, corporate governance measures should include those nonfinancial aspects as well.

Globalization has opened new avenues for organizations to expand in multiple directions, thus markets are better managed, and governed organizations have a higher probability of success as they may seem more attractive to foreign markets. Even for organizations working in local markets, corporate governance plays a significant role in setting up the operating structure and policies, which results in higher profitability.

An increased awareness by the general public has heightened the expectations from organizations, and they are expected to do more for the betterment of people and society based on their location. There seems to be an understanding that organizations are part of the social system and should not focus solely on their profitability; rather, organizations should address the concerns of related constituents as well. Thus, the social responsibility perspective has been added to the corporate governance mechanism.

Availability of technology has increased information availability and communication frequency between all stakeholders. This has caused organizations, especially their executive management, to always be in a state where they have ensured that precise controls are in place to safeguard the interests of the stakeholders. Efficient and aligned corporate governance practices can help organizations achieve this control.

The Sarbanes–Oxley Act of 2002 (SOX) and other regulations have had a major impact on the way that organizations are being governed. Some of these regulations are mandatory in certain countries; however, organizations tend to adapt to other codes and standards as well, due to the belief that such adaptation will result in the establishment of goodwill in the market. However, laws and regulations may have loopholes, which can be exploited. Corporate governance creates mechanisms of self-regulation and reduces dependency on state-level legal options, which are not always in the best interest of the corporation.

Recent corporate debacles and scandals have undermined the confidence of investors and questioned the ability of organizations to protect the interests of their stakeholders. As a result, the focus of organizations as well as regulators shifted to the tightening of regulatory requirements, and revisions in corporate governance frameworks have started to take place. According to the OECD, the following factors have an influence on the evolution and implementation of a governance framework within an organization (OECD 2004):

1. Macroeconomic policies of the environment
2. Degree of competition in the product and factor markets
3. Legal and regulatory requirements
4. Institutional culture and environment
5. Business ethics prevalent in the society within which the organization is operating
6. Corporate awareness and interest that the society is taking in the organization
7. Relationships among participants of the corporate governance framework, their influence, and their special demands, for example, investors wanting a certain reporting mechanism to be in place.

CORPORATE GOVERNANCE INSTITUTIONS

The Board of Directors

The board of directors acts as the main driving force within the corporate governance framework. The board's performance has implications on the way corporations operate as well as perform. Recent tragedies, such as the British Petroleum (BP) oil spill in 2010, required organizations to reexamine and evaluate corporate governance practices and performance. The focus should now be on improving board performance to provide better accountability and transparency in order to enhance corporate performance. Setting up metrics to measure board performance is another key aspect that corporations should consider. This will provide a quantitative assessment tool to evaluate how the board is contributing to the overall board objectives.

The board of directors can be seen as a controlling entity, which is formed by the shareholders and their nominating committee to oversee the activities of the management team. The presence of independent directors on the board may result in lowering the probability of self-seeking behavior of the management team.

Boards play a vital role in reviewing, recommending, and approving strategic plans. They have the responsibility of ensuring that the organization is investing in the right direction and that the investments are generating the desired results. Corporate accountability is another board-level function, and boards have to ensure that the company discloses relevant and reliable financial and nonfinancial information to various internal and external stakeholders. The corporate board has to create a working environment that promotes governance, transparency, and ethical behavior within the corporation. Boards are also responsible for senior level hiring, compensation, and succession planning.

Effective and reliable management reporting improves as the board performs, and accurate reports result because of the facilitation of board-level decisions. In one of my recent assignments, while performing the as-is analysis, it was clear that the board members were obtaining information that was irrelevant, presented ineffectively, or overloaded with details. This approach means that board members make decisions based on their understanding of information, which may not be complete or wholly correct. Therefore, it is important that reports be timely, accurate, and presentable, and they should also contain quality information.

In terms of board member selection, the independence of the board of directors seems to be a logical preference as they will act as a monitoring function over the management team. There appears to be a conflict of interest if the CEO plays a significant role in the selection of board members, as the selected board members might be expected to reciprocate the favor in some form. That might be the reason why the SOX Act has assigned more responsibility to the nomination committee for board member selection.

There is, however, the possibility that even after the selection, the independent board member might not want to alienate senior management and will try to work out suboptimal solutions, especially if they feel that their reelection to the board might be hindered. We can thus argue here that boards should not be considered as the only option when designing control frameworks, and complementary approaches should be used in conjunction with a focus on corporate social responsibility as well.

Auditors and Their Independence

Auditors play an important role in the corporate governance mechanism. External auditors tend to play a dual role, that is, partners to the board and independent monitors. On the one hand, they work as partners to the board of directors to ensure that the management team, in particular, and the organization, in general, is complying with standards and regulations. On the other hand, they may report issues, which if made public, can result in organizational loss.

Standards and regulations require the appointment of external independent auditors. Researchers and practitioners have, however, questioned the independence of external auditors because of the assumption that external auditors will always remain unbiased while reviewing the client's reports and books. This bias would be created perhaps because of financial incentives as well as nonfinancial reasons, such as personal affiliation, which may be due to a long-term association with the client. Bias is not a synonym for fraud, but it affects the auditing process as external auditors might ignore certain inconsistencies in the reports. However, such ignorance on a consistent basis might present an opportunity for a self-seeking opportunist to commit irregularities. Marnet (2005) mentions certain mandatory requirements that may help in reducing such bias and mutually supporting behavior.

1. Mandatory rotation of external auditors.
2. Distance and anonymity between an organization's employees and external auditors.

FUNCTIONS OF CORPORATE GOVERNANCE

Reporting and Disclosure

Efficient, effective, and comprehensive reporting is at the core of corporate governance. A governance mechanism can have its strategic value in the reporting system, and the information contained within the reports. The objectives of transparency and full disclosure, which are important to stakeholders, can truly be achieved through timely, comprehensive, clear, and accurate reporting.

Resource Management

Resource management as an activity is a management level responsibility; however, corporate governance should set up the framework under which management makes the investment decisions. Also in certain cases, such as executive level hiring, the corporate governance institutions, such as the board of directors, act as an approving authority.

Risk Management

Risk management is a key attribute of corporate governance. Effective risk management helps organizations to mitigate potential risks to an acceptable level. Without an effective risk management framework, corporate governance is viewed by the market as risky. Therefore good governance, which includes risk management as a practice, results in instilling a sense of belief in the creditors and investors.

Performance Management

Performance of the organization should be a key function of the governance framework. Performance is measured in terms of effectiveness and efficiency in maximizing the value for all stakeholders with a special focus on shareholders. Governance should be carried out beyond the financial indicators. Thus, performance is not just about financial performance, even though it matters the most. It is also related to nonfinancial matters such as effective human resource management and organizational process efficiency.

Relationship Management

Managing the relationships with all stakeholders is a governance responsibility. Pultorak (2005) calls this *relating responsibility* and states that the governance function should be there to work with the stakeholders, pay attention to their needs, manage expectations, and balance their requirements from the corporation.

Corporate social and ethical responsibility to society should also be considered. This function of governance is related to all the other functions, as one of the major reasons for the other governance focus is to sustain and build a clear and transparent relationship with the shareholders, in particular, and all stakeholders, in general.

Strategic Oversight

Strategic oversight is a major function of corporate governance. The development and implementation of strategic plans is a management responsibility; however, corporate governance institutions, such as corporate boards, have a major role to play in terms of reviewing, approving, and overseeing the implementation of strategic plans.

Compliance

Compliance is a concept that deals with how organizations work under a framework of principles, values, policies, and codes. Regulations, such as SOX, focus on accountability of the executive management and the board of directors. Such regulations also focus on the independence of auditors and disclosure of financial and nonfinancial reports to all stakeholders. Compliance with such standards helps in reducing the probability of issues related to fraudulent and opportunistic behavior of the executives. Compliance with these standards is no longer a matter of choice, and organizations have to adapt if they want to operate, especially in progressing markets. Pultorak (2005) refers to this dimension as conformance to legislative requirements. Legal compliance, which is related to complying with standards and codes, should go hand in hand with ethical compliance, which is concerned with the values and norms of people.

THE HUMAN SIDE OF CORPORATE GOVERNANCE

Most of the perspectives on corporate governance seem to focus on what is to be done without concern for the human side of this function. This may be because of the complexity involved in dealing with the prediction of human behavior. However, these perspectives do help in understanding the decisions that organizations make, which may not make sense to some stakeholders.

Ethics and Ethical Compliance

The focus of corporate governance has moved from shareholder interests to stakeholder interests. These stakeholders exist in the environment and the society within which the corporation is operating. This has resulted in corporate social responsibility and business ethics becoming an integral part of the corporate governance framework.

Legal compliance frameworks have proven to be insufficient as they lack a moral foundation, which is necessary for people and corporations to behave in a responsible manner. I read a research paper by Arjoon (2005), and I can truly align with his argument that "The current environment of failures of corporate responsibility are not only failures of legal compliance, but more fundamentally failures to do the right (ethical) thing" (p. 343). We should not disregard the importance of legal compliance; however, there should be a balance created between the legal and ethical aspects of compliance.

In his book, Collins (2001) emphasized the importance of ethics, character, and values. Level 5 leaders in companies that showed sustained growth and outperformed the market displayed a balance of personal humility and professional will. These leaders were ambitious and showed diligence, however, their effort and ambition were for the growth of the organization and the stakeholders. Level 5 leaders were also people of integrity and conscience, who were ready to take the blame when things went wrong. They emphasized getting the right people on the board, where "right" has more to do with character traits and instinctive capabilities instead of specific knowledge.

Obedience to a law sometimes becomes such an overwhelming concept that people tend to forget why that law actually exists. This concept of legalism results in an environment where people tend to do things because "they have to do them" not because "they want to do them." The end result is an organizational culture which does not necessarily promote excellence but rather focuses on compliance. People should not simply follow what is there in the book; instead organizations should focus on developing an ethical culture where people are governed by common values and principles. They should really believe in the laws and should agree to follow them because "they want to" not because "they have to." This instills a culture of excellence where laws and ethical behavior support each other.

Collins's research on companies that display consistent excellence and growth validates this concept when he states that the "right people" play an important role in creating a culture of self-discipline as they take disciplined actions without any enforcement. Such cultures have a dual facet, which means that people work under consistent systems and processes; however, the environment provides them with an opportunity to be responsible, creative, and excel within the system's framework. Such a system ejects people who do not share the same values and standards of the organization.

Following corporate governance practices based on core values of integrity and trust is not just about doing the right thing. It may result in attracting and retaining talent (employee loyalty) and building goodwill in the market, which in turn may increase the profitability of the organization. Also, whenever such organizations face situations where the law is ambiguous there is a high probability that the employees will make the right choices. Collins's "good to great" organizations displayed this integrity and discipline as one of their main traits. Arjoon (2005, p. 349) summarizes this by mentioning that all organizations have the right to make profits and grow (that is why they are in business), however, "the pursuit of profits must stay within ethical bounds."

Rationality and Behavioral Aspects

Cognitive and behavioral attributes of bounded rationality and opportunism play a significant role in how people behave under different circumstances and how they make decisions, which may not seem to be rational to others. Governance can be considered as a behavior that defines the relationship between a firm and its stakeholders. Auditors and boards of directors are, mistakenly, expected to behave in a rational way and care about their reputation. However, being human, they are bounded by human traits, where they may care more about immediate gain without sufficient regard for potential negative future consequences. Certain illogical behaviors shown by the gatekeepers, such as sacrificing their reputations for small gains, validates the presence of bounded rationality and opportunist behavior.

Behaviors and bounded rationality of different actors can be due to their background and knowledge; however, other factors such as context and situation also play a significant role in decision making. Marnet (2005) considers that decision making under uncertainty is influenced by "biases, schemata, anchoring, framing, and other cognitive and judgmental shortcuts of the human mind" (p. 619).

Careful consideration must be paid to performance criteria while setting up performance-based incentive mechanisms. This is because in certain situations performance incentive schemes for the management team may result in the downfall of the organization as it may result in internal conflicts and unethical behavior especially when people are pursuing the same goal and are just short of the target.

DESIGN AND IMPLEMENTATION CONSIDERATIONS

Effective implementation of corporate governance depends on various factors such as legal, social, economic, and organizational context.

Establishment of the governance regime is the responsibility of senior executive management, however, certain constraints affect this establishment such as:

1. Factors which are external to the organization's direct control such as laws of the country for taxation.
2. Sector-specific factors, such as regulatory requirements of the industry within which the organization is operating.
3. Organizational culture and environment.

Factors such as these and others should be taken into account while forming a suitable governance regime (OGC 2011).

For effective governance, the governance mechanism should be aligned with desirable behavior, and must be updated when the desirable behaviors change. I first came across the term *desirable behavior* while reviewing the work of Weil and Ross (2004). Desirable behaviors are those behavioral aspects of employees that are aligned with the organization's values and principles. Desirable behaviors differ from organization to organization and are of prime importance to effective governance. Desirable behaviors define the expected behavior from the organizational participants and are voiced through various sources such as vision, mission, values, and organizational principles.

While referring to the CPR framework, Pultorak (2005) mentions that all three areas, namely, conformance, performance, and relating responsibility, are important when designing a governance framework. These should govern four assets: infrastructure, value creation, internal processes, and people; and clients and external stakeholders. He mentions that focus on only one aspect will not result in optimal value from governance. This means that, if an organization is merely focusing on conformance, it will surely comply with all regulatory and legislative requirements; however, there is a strong possibility that the future growth of the organization will be hindered by this approach as there will be no focus on performance aspects as well as relationship building with customers and other stakeholders.

In recent years, a focus on standards and codes has increased, and new codes have been developed, which tend to control and guide organizations in the implementation of corporate governance mechanisms. This is due to the reason that the presence of laws, which are enforceable through the legal system, are not sufficient, and in some cases inefficient, in the implementation of governance frameworks within corporations. These standards align to best practices as well as the legal system. Thus, adherence to these codes has a dual benefit. Implementation of the standards leads to development of self-regulatory principles, as well as adaptation to international standards.

The OECD principles provide guidelines for what should be considered when designing a corporate governance framework (OECD 2004). Based on the OECD's recommendation the governance framework should:

1. Promote transparent and efficient markets, be consistent with the rule of law, and clearly articulate the division of responsibilities among different supervisory, regulatory, and enforcement authorities.
2. Protect and facilitate the exercise of shareholders' rights.
3. Ensure the equitable treatment of all shareholders, including minority and foreign shareholders. All shareholders should have the opportunity to obtain effective redress for violation of their rights.
4. Recognize the rights of stakeholders established by law or through mutual agreements and encourage active cooperation between corporations and stakeholders in creating wealth, jobs, and the sustainability of financially sound enterprises.
5. Ensure that timely and accurate disclosure is made on all material matters regarding the corporation, including the financial situation, performance, ownership, and governance of the company.
6. Ensure the strategic guidance of the company, the effective monitoring of management by the board, and the board's accountability to the company and the shareholders.

An integrated communication environment must be created between all entities of the corporate governance function. This means that the mechanisms of communication, the protocol, and the medium of communication should be standardized and well defined. This will result in a standard organizational vocabulary resulting in fewer number of issues that are due to communication gaps.

SUMMARY

Failure to perform effective corporate governance practices can have a negative impact not only on the organization but also to the society within which it is operating. Investors and creditors may lose confidence in the organization, and the organization may face social as well as legal pressures from society and the legislative system, which may result in loss of business.

Organizations, such as OECD, around the world are continuously developing and updating codes and principles related to corporate governance to meet the ever-changing needs of the markets. Thus, organizations that want to remain socially acceptable and profitable have to adapt and undergo continuous reforms in their governance frameworks.

While designing the governance framework, attention must be paid to the impact it will have on the market within which the organization is operating and its shareholders and stakeholders. The rights of all stakeholders must be defined; thus, the concept of corporate social responsibility must be brought into the picture. However, special attention must be paid to the shareholders. Disclosure of information transparency and efficient reporting should also be a major focus area. Corporate governance frameworks should also clarify the responsibilities, authority, and accountability of all stakeholders involved in the governance mechanisms, most importantly the board members.

Corporate governance frameworks should allow for human behavior and attributes. Too much reliance on processes, independent monitors, and structures, with the assumptions that humans will always behave rationally, has resulted in corporate disasters. Marnet (2005) questions "whether a dependence on rules and regulations that ignores human ability to rationalize actions has the intended deterrence effects" (p. 628).

Corporate governance practices help organizations to grow and progress within a controlled framework. Organizations following credible corporate governance practice have the confidence that they are not only doing

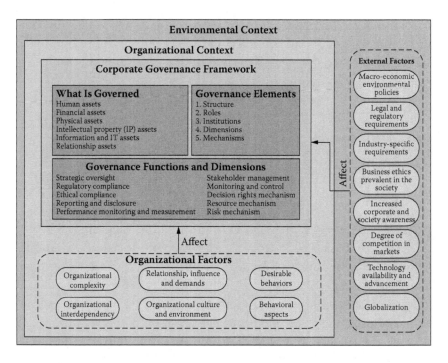

FIGURE 4.1

Corporate governance—conceptual model.

the "right thing" they are doing it in the "right manner." The outcome of this is that the outside investors have more confidence, and governed organizations have a lower risk profile and fewer unexpected results, thus the financing sources become more stable for such organizations and markets. Good corporate governance also results in a more organized operational mechanism at the organizational level, and the markets within which such organizations are operating also obtain a positive impact resulting in improved economic growth.

The diagram in Figure 4.1 provides a consolidated view of corporate governance.

Before focusing on program governance, it is essential to develop a thorough understanding of a program and its related concepts. The following chapters focus on the underlying concepts, as well as practical, organizational, and operational aspects, related to programs.

REFERENCES

Arjoon, S. (2005). Corporate governance: An ethical perspective. *Journal of Business Ethics,* *61*(4), 343–352.

Chau, S. L. (2011). An anatomy of corporate governance. *The IUP Journal of Corporate Governance, X*(1), 7–21.

Collins, J. (2001). *Good to Great.* London: Random House Business Books.

Marnet, O. (2005). Behavior and rationality in corporate governance. *Journal of Economic Issues, 39*(3), 613–632.

Office of Government Commerce (OGC). (2011). *Governance.* Retrieved 10th April 2011, from http://www.ogc.gov.uk/delivery_lifecycle_governance.asp.

Organisation for Economic Co-operation and Development (OECD). (2004). *OECD Principles of Corporate Governance.* Paris, France: OECD Publishing.

Pultorak, D. (2005). IT governance: Toward a unified framework linked to and driven by corporate governance. In *CIO Wisdom II* (pp. 283–320). New Jersey: Prentice Hall.

Weil, P. and Ross, J. W. (2004). *IT Governance: How Top Performers Manage IT Decision Rights for Superior Results.* Boston, MA: Harvard Business School Press.

Section II

Programs

5

Temporary Organizations

WHAT ARE TEMPORARY ORGANIZATIONS?

Temporary organizations have been in existence for centuries; however, their existence in literature can be traced back to the 1960s and 1970s. The definition of temporary organizations differed based on the different perspective of the researchers.

Most of the work that I have done was in situations where we had to organize, plan, execute, and deliver results within a predefined timeline and other constraints. Palisi (1970) called such organizations *transitory organizations*, and stated that such organizations "are intended to be organized for a short span of time or for a limited number of meetings" (p. 200). Permanent organizations, which cease to exist because of various factors, cannot be called *temporary* as temporariness was not planned for such organizations. Two main attributes of a temporary organization identified by Palisi (1970) are:

1. Planned or intended temporariness
2. Immediate goal achievement

Miles (1977) referred to them as *temporary systems* and explained that temporary systems are a "time-limited system, to be terminated by advance agreement when certain states, events, or points in time have been reached" (p. 134). One of the primary attributes of such systems is that they are vehicles or tools to induce change in the permanent systems since they can focus on a certain objective(s) at hand, drive actions, and deliver results that cannot be delivered by existing structures.

Goodman and Goodman (1976) took an internal perspective and defined temporary organizations as "a set of diversely skilled people working together on a complex task over a limited period" (p. 494). Packendorff (1995) also took an insider's view and stated that temporary organizations have the following four traits:

1. A set of collective actions that initiates a nonroutine process and/or produces a nonroutine product
2. Are time bound
3. Have predefined performance expectations
4. Need a conscious organizing effort because of their complexity

This diversification in definition may be due to the number of different forms that temporary organizations take based on the environment they are operating in, or it may be due to the approaches that researchers have taken while explaining the concept. This diversity also proves that the concept of temporary organizations is not limited; rather it is being applied in various industries in various forms and situations.

Two key factors within temporary organizations are the concepts of time and action. Normal organizations are formed to last for long durations and are more focused toward decisions that will have a long-term impact. Temporary organizations do not have this luxury of time and are more concerned about taking actions based on the decisions made or actions originating from the surrounding circumstances for which decisions are made to justify the action (Lundin and Söderholm 1995). This focus on action is due to the temporary nature of such organizations, as results have to be given within the time constraint. One key point that should be mentioned here is that temporary organizations are time bound, that is, they have a predefined termination point; however, the results produced by these organizations may outlive this temporariness and may have a long-lasting effect. Lundin and Söderholm (1995) propose four attributes of temporary organizations that separate them from nontemporary organizations. They say that temporary organizations have an:

1. Objective or a task to accomplish, by a
2. Team, within a
3. Defined period of time, and that at the end there should be a
4. Transition from the current to a future state.

While referring to the concept of sequencing and life cycle in a temporary organization, Lundin and Söderholm (1995) specify four distinct phases:

1. Action-based entrepreneurialism, which is related to identifying the need, building the business case, and gathering the necessary pressure to initiate the temporary organization.
2. Fragmentation for commitment building, which is related to planning for the tasks, timelines, closure, and other related areas for the temporary organization. It also includes commitment building, from all members, to the agreed-upon timelines, activities, and other plans. This fragmentation is achieved through time bracketing, which includes officially announcing the commencement of the project/program in order to decouple the project/program from its surrounding past, contemporary and future activities, and task partitioning, which mentions that the objective and the related tasks of the projects/programs have to be spelled out to ensure that such tasks are delimited to exclude out-of-scope tasks.
3. Planned isolation is related to the execution phase of the temporary organization's life cycle. Here the focus is to ensure that the plan gets executed in order to achieve the desired results. This is achieved through deliberately isolating the temporary organization from its surroundings so as to minimize external influences and threats.
4. Institutionalized termination relates to the dissolution of the temporary organization. This includes activities of recoupling the temporary organization with its surroundings as well as transmission of knowledge and experiences from the temporary organization to the production or operations functions. This recoupling, which can be time driven, event driven, or condition linked occurs after the temporary organization ceases to exist.

Whitley (2006) mentions temporary organizations as entities that are set up to perform certain tasks and are terminated upon completion of the task. He further differentiates temporary organizations, mentioned as project-based firms (PBF), based on singularity of task along with uniqueness and stability of roles, skills, and responsibilities.

It is obvious from the above that there has been a great deal of attention paid to the concept of temporary organizations in recent years. This is due to the understanding that they have a predefined termination point, and thus are a crucial form of a modern-day economic organization (Kenis et al. 2009). One can also observe that the researchers above have focused on temporary organizations from a project or program perspective. However, temporary organizations encompass a much greater domain, which has projects and programs as one of its subdomains. Kenis et al. (2009) suggest that "temporary organizations are a conceptual category that encompasses projects but also other forms of temporary organizing" (p. 60).

PROJECTS AND PROGRAMS AS TEMPORARY ORGANIZATIONS

When we look at projects as temporary organizations instead of projects as a tool the focus shifts from plan, control, and evaluate to expectations, actions, and learning (Packendorff 1995). Turner and Müller (2003) took an agency perspective, and viewed projects as temporary organizations based on the following:

1. As a production function
2. As an agency for assigning and utilizing resources in order to manage beneficial change in an organization
3. As an agency to manage uncertainty that is part of such changes

Turner and Keegen (2001) define the project-based organization as one in which the "majority of products are made or services are against bespoke design for customers" (p. 256). While taking a transaction cost economics perspective and equating projects with transactions, they talk about undertaking different types of organizational projects and elaborate on different scenarios that project-based organizations assume. Even though such organizations are permanent in nature, they manage or undertake multiple goals through different temporary organizations. They divide project-based organizations based on whether:

1. They undertake multiple projects for multiple clients, which is basically a portfolio of projects and which can be temporary if the portfolio was created to achieve certain strategic objectives through multiple projects and programs.
2. They undertake a program of projects for customers in which all projects are in the end contributing to the program objectives. This can be considered a type of temporary organization.

3. They undertake large projects for large clients such as large construction initiatives. This is a categorization of projects based on magnitude and is also a temporary organization.
4. They are a start-up company created to develop a product or service. They consider that the company and the product it is creating is a project on its own. We can treat such start-up companies as temporary organizations if they are formed with a mind-set that they will be terminated once the product is developed, or the service is delivered. However, if they are created with an objective to exist in the market then the project that creates the product can be treated as a temporary organization, however, the organization cannot be called *temporary*.

Certain organizations call ongoing operational activities *programs*. The Project Management Institute (PMI®) (2013) also mentions that "Some organizations and industries refer to ongoing or cyclical streams of operational or functional work as programs" (p. 1). Management and governance aspects of such programs come under the domain of operations or general management as such organizations can be considered as permanent setups in large organizations.

SUMMARY

In summary, temporary organizations are viewed as organizations which are formed to achieve some sort of major objective and are disbanded after:

1. The objective is achieved.
2. The objective no longer remains viable because of internal or external circumstances.

The word *temporary* in temporary organizations itself implies the time limitation of such organizations. At the end of the life cycle, temporary organizations terminate by either becoming part of the normal operations or transitioning the achieved objectives or benefits to another entity. Thus, the three main attributes as depicted in Figure 5.1, unique to temporary organizations that have been derived from the above are:

FIGURE 5.1
Common attributes of programs and temporary organizations.

1. Time bound
2. Goal/objective achievement
3. Transition

The rest of the attributes related to a temporary organization can also be traced back to other organizational forms, which may not be temporary in nature.

Permanent organizations are focused toward continuity and long-term growth; however, projects and programs are focused toward achieving strategic goals or objectives and cease to exist based on predefined conditions. Thus, projects and programs can be viewed as temporary organizations within the permanent organizational framework. The main focus of this book are programs that come under the domain of temporary organizations and they are discussed in the next chapter.

REFERENCES

Goodman, R. A. and Goodman, L. P. (1976). Some management issues in temporary systems: A study of professional development and manpower—The theater case. *Administrative Science Quarterly, 21*(3), 494–501.

Kenis, P., Panjaitan, M. J., and Bakker, R. M. (2009). Research on temporary organizations: The state of the art and distinct approaches towards "temporariness." In Kenis, P., Panjaitan, M. J., and Cambré, B. (Ed.), *Temporary Organizations Prevalence, Logic and Effectiveness.* Cheltenham: Edward Elgar Publishing Limited.

Lundin, R. A. and Söderholm, A. (1995). A theory of the temporary organization. *Scandinavian Journal of Management, 11*(4), 18.

Miles, B. (1977). On the origin of the concept "Temporary Systems." *Administrative Science Quarterly, 22*(1), 134–135.

Packendorff, J. (1995). Inquiring into the temporary organization: New directions for project management research. *Scandinavian Journal of Management, 11*(4), 319–333.

Palisi, J. (1970). Some suggestions about the transitory-permanence dimension of organizations. *British Journal of Sociology, 21*(2), 200–206.

Project Management Institute (PMI). (2013). *The Standard for Program Management* (Third Edition). Newtown Square, PA: Project Management Institute.

Turner, J. R. and Keegan, A. (2001). Mechanisms of governance in the project-based organization: Roles of the broker and steward. *European Management Journal, 19*(3), 254–267.

Turner, J. R. and Müller, R. (2003). On the nature of project as a temporary organization. *International Journal of Project Management, 21*(1), 1–8.

Whitley, R. (2006). Project-based firms: New organizational form or variations on a theme? *Industrial and Corporate Change, 15*(1), 77–99.

6

Programs

Is a megaproject a program? I had this question in mind for years; however, I was unable to get a clear answer to this question. Some researchers and organizations considered megaprojects to be under the program domain, whereas others considered them as projects. During one of my visits to SKEMA University in France, I had a discussion on this subject with Rodney Turner, an authority in the field of project management. After a long conversation, we concluded that megaprojects have attributes that are more aligned with programs than with projects, and that they can be studied under the program management domain.

For now, I align myself with the understanding that megaprojects can be investigated under the domain of programs, but this is an area which can be considered blurry at best. However, one established fact is that projects and programs are considered as vehicles, or tools, that drive the organization toward the achievement of strategic goals.

DEFINING PROGRAMS

Organizations tend to organize projects in a way so that the probability of realizing the organizational strategy and consolidated business benefits is increased. In such cases, projects do not exist in isolation, as there are interdependencies between them which have to be resolved and managed. While an individual project's emphasis is on achieving objectives related to it, the overarching program focus is on managing activities outside the project and the interdependencies between composite projects. The management of these aspects is critical to realize the benefits expected from the interdependent projects. We could consider a program as a vital link between organizational strategy and its execution through projects.

From an external perspective, programs are considered as governance structures that are designed to manage and coordinate related projects. Programs are focused toward achieving higher-level organizational objectives through a collection of projects each of which contributes to specified objective(s). Projects deliver outputs/outcomes, which combine to provide the program level benefits.

The definition of a program has evolved over the years. Gray (1997) defined programs as "A group of related projects which together achieve a common purpose in support of the strategic aims of the business." One of the core items missing from this definition was the aspect of coordinated management of projects, their interdependencies, and the undertaking of nonproject activities that are within the scope of the program. Pellegrinelli (1997) added the aspect of coordinated management, and defined a program as

> A framework for grouping existing projects or defining new projects, and for focusing all the activities required to achieve a set of major benefits. These projects are managed in a coordinated way, either to achieve a common goal or to extract benefits which would otherwise not be realized if they were managed independently. (p. 142)

As the evolution continued, researchers realized that programs act as a vehicle to deliver high-level strategic objectives. This aspect was added by Murray-Webster and Thiry (2000) when they defined a program as "a collection of change actions (projects and operational activities) purposefully grouped together to realize strategic and/or tactical benefits" (p. 48). This concept is also echoed by Turner and Müller (2003), as they mention that programs are generally focused toward higher-level strategic goals, which are generally long term and less specific, and projects contribute, through their outcomes, in the achievement of these strategic goals. Turner and Müller's (2003) definition states that a "program of projects is a temporary organization in which a group of projects are managed together to deliver higher order strategic objectives not delivered by any of the projects on their own" (p. 7).

The definitions were later refined by project management bodies, and OGC (2009) referred to programs as a "temporary flexible organization structure created to coordinate, direct and oversee the implementation of a set of related projects and activities in order to deliver outcomes and benefits related to the organization's strategic objectives. A program is likely to have a life that spans several years" (p. 309).

APM (2009) contributed by defining a program as it "is a group of related projects, which may include related business-as-usual activities, that together achieve a beneficial change of a strategic nature for an organization" (p. XV).

PMI, in its third edition of the program management standard, refined the definition of program and added the aspect of subprograms. PMI (2013b) defines a program as "a group of related projects, subprograms, and program activities that are managed in a coordinated way to obtain benefits not available from managing them individually" (p. 4).

While reviewing the definitions of institutions, we can see that OGC (2009) refers to the longevity, temporariness, and flexibility of programs; whereas PMI (2013b) and APM (2009) additionally mention that the program may include work outside the scope of individual projects. All of these definitions, however, agree that the projects or subprograms within the program are related and are contributing to certain shared strategic objective(s).

PROGRAM CHARACTERISTICS

Based on the definitions discussed above, certain specific characteristics of programs are identified in Figure 6.1, and discussed below.

Benefits Delivery

Programs focus on creating and delivering benefits by the coordinated management of projects and related activities. Individual projects focus on delivering specific output or outcomes. Programs focus on combining these outcomes to deliver strategic objectives.

The delivery of these objectives results in benefits that have to be realized by the beneficiaries. The benefits of the program may be delivered during the program execution, as projects deliver results, at the end of the program life cycle, or years after the actual deployment of the program deliverables. Examples would be internal restructuring programs that

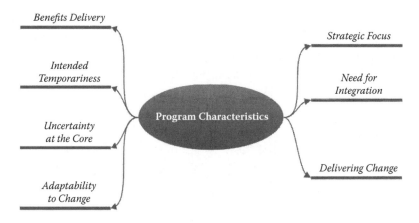

FIGURE 6.1
Program characteristics.

deliver benefits throughout their execution or Burj Khalifa (the world's tallest building in Dubai) that delivered benefits after its completion and is continuing to do so.

Strategic Focus

Programs are considered vehicles to execute corporate strategy and delivery benefits to the organization. According to PMI (2013b), "Programs are designed to align with organizational strategy and ensure organizational benefits are realized" (p. 25). Thus, we can consider a program as a link between strategic plans and execution through composite projects and related activities. This understanding is verified by Artto et al. (2009) when they state that programs have a greater strategic focus as compared to projects. As the organizational strategy alters, the associated programs need to adapt accordingly.

From another perspective, the results of the program may also impact the organizational strategy. I was heading a large change management program at a Gulf Cooperation Countries (GCC) government entity. The core objective of the program was to implement processes and redesign the organizational structure to improve the overall capability and delivery capacity of its IT department. Through realistic planning and careful execution, the program was able to deliver some of its intended benefits three months before the agreed timelines. This early delivery triggered the department's governance board to approve and initiate two new projects with objectives that were related to process automation.

Intended Temporariness

Programs, as per PMI, are temporary organizations. This characteristic of transience is considered to be the most important and has an influence on other characteristics. However, the most important aspect of temporariness is the "intention of temporariness" from the beginning; otherwise all organizations, on some time scales can be considered as temporary in nature, whereas some organizations might not be so temporary, meaning that they may last for durations longer than expected.

The Need for Integration

Programs create a need for integration between different parts of the program as well as the integration of the program with the organization. In case of projects, OGC refers to this as cross-functionality, that is, projects require cross-functional interaction within organization units and between people, belonging to different trades and skills, who work together, as a team, to deliver the project objective (OGC 2009). However, when it comes to the level of a program there is a need for integration between the projects of a program as all of them working toward delivering the same benefit(s).

The program also needs to integrate the delivered benefits to the beneficiary organization. This is required because the realization of benefits can only be achieved if the beneficiary entity can integrate the benefits as part of its routine operations and sustain them.

Uncertainty at the Core

Programs address organizational challenges, which cannot be handled through routine operations. Such challenges have uncertainty built at their core, making programs riskier and uncertain. This attribute also exists in projects. However, projects are considered more certain than programs. PMI (2013b) states that the expected outcomes of a project are relatively clear at the time of its inception. This clarity is relatively low when it comes to programs.

The focus of a program on benefits delivery and strategic alignment, enforces the program team to adjust the direction of the program during its life cycle to ensure that it achieves the desired benefits. New projects might have to be initiated, existing projects might be canceled, and the overall road map of the program may be revised to ensure alignment with the intended benefits delivery. All this is part of the uncertainty aspect that exists in programs.

Thus, it is important for the program management and governance team to understand this core characteristic. They need to ensure that the program's management, governance, and decision-making mechanisms take uncertainty into consideration as a key factor.

Delivering Change

Programs act as an agent to induce change in the organization; however, the magnitude of change is much greater than what one would expect from a project. Depending on the type of program, it may deliver incremental benefits, which will be transitioned to the beneficiary organization. There might be other programs that deliver their benefits after completion of all component projects, at the program closure.

Adaptability to Change

Programs respond to changes with an entirely different approach as compared to a project. Projects apply a conservative approach when responding to changes. The goal of the project manager is to ensure that changes to the plan are minimized. The strategic alignment and benefits delivery focus of programs dictate that programs embrace change in order to keep them aligned with the context within which they are executing. The highly uncertain nature of the program also influences the program management and governance team to adapt, based on changing internal context, such as delivery and quality of project outcomes and their integration and external factors, such as changes in legislation or evolving organizational priorities.

PROGRAM CATEGORIZATION

Organizations tend to categorize programs based on different characteristics. This categorization seems to be logical since the uncertain nature of the initiatives, along with the environment in which they are executed, dictates that the management and governance mechanism needed by these initiatives should vary based on program attributes.

PMI (2013b) classified programs on the basis of a delivery model. Certain programs deliver benefits at different stages of their life cycle that can be leveraged by the beneficiaries. Such programs require a consistent mechanism to transition the incremental benefits to recipients, and the recipient entities should be ready to leverage and sustain the benefits delivered. The focus here is the incorporation and transition of benefits, at the recipient organization, as the program unfolds and delivers.

Examples of such programs are IT infrastructure implementation programs where certain components of the infrastructure are delivered to the organization for utilization, whereas other components may be delivered at a later stage. Another example would be organizational capability enhancement programs that include the implementation of processes and supporting tools. Such an initiative will roll out developed processes and tools in a staged fashion so that the organization can leverage the immediate benefits, and at the same time increase its maturity level so that it can handle more complex deliverables at a later stage.

Programs, such as major construction initiatives, medical research, or aerospace, deliver benefits at the end of the program life cycle, that is, when all the component projects have been completed. Such programs require a greater need for integration, as the final benefit(s) is directly dependent on the outcome of the component projects. In such cases, the type of planning required to deliver beneficial change to the organization might be entirely different as compared to programs delivering incremental changes. The focus here will be more on integration and recipient readiness.

The Global Alliance for Project Performance Standards (GAPPS) conducted a major research initiative and classified programs based on three criteria (2011):

1. The type of benefit(s) delivered.
2. Interdependence between constituent projects and their alignment with strategy.
3. The objective of grouping them under one program.

GAPPS (2011) classified programs as:

1. Strategy programs that are focused on delivering benefits that impact the organization's immediate future state. The constituent projects are tightly linked with a specific strategic objective. In addition, the

benefits delivery mechanism of such programs are iterative such that results on initial outcomes impact the decisions that follow.

2. Operational programs that deliver benefits that positively impact the ongoing operations of the organization. Such programs have a close interdependence between the constituent projects. To minimize the impact on ongoing operations such programs also deliver benefits in an incremental manner.

3. Multiproject programs, which are groupings of projects that share similar customers or technology or are grouped to achieve synergies by increasing efficiency in terms of resource utilization. The interdependence of projects in such cases is minimal. PMI (2013b) classifies such groupings as portfolios rather than a program.

4. Megaprojects that deliver a specific asset to the organization. These are basically projects that are of sizes significantly larger than the typical projects handled by the organization. If we align ourselves to the PMI (2013b) terminology, megaproject will still conform to the definition of project.

The GAPPS (2011) classification is partially based on earlier research conducted by Pellegrinelli (1997) where three program types were identified. Pellegrinelli (1997) classified programs as:

1. Portfolio-type programs that have relatively independent projects and are grouped together based on a shared theme such as using a common resource or skill base. This coordinated management enhances the utilization of resources or improves the chances of project delivery.

2. Goal-oriented programs that are initiated to execute a strategic decision, which cannot be executed within the framework of routine operations. Such programs act as a vehicle to take organizations from one state to another.

3. Heartbeat programs that are initiated to enhance the current capabilities, processes, or systems in an organization. The objective of such programs is to ensure continuous improvement in the organization's delivery or operational capacity. Such programs tend to incrementally integrate the benefits to the daily operations with minimal disruption.

Pellegrinelli's (1997) classification was also analyzed by Vereecke et al. (2003) in order to define a typology of programs based on two key factors:

1. The extent to which projects already exist at the time of program formulation.
2. The significance of the impact of program benefits on the beneficiaries.

They classified programs as follows:

1. Type A programs that have a large number of projects at the outset and are designed to bring gradual changes in the recipient organization.
2. Type B programs that have a large number of projects which are integrated to deliver a significant change at the recipient organization.
3. Type C programs that start as a new initiative and have an objective of enhancing a current organizational capability.
4. Type D programs that start as a new initiative and have an objective to add a new capability at the recipient organization, that is, employing a significant change.

A typology of programs has also been developed by OGC (2009) where the programs are divided into vision-led, emergent, and compliance types, each having its own specific attributes.

Another important aspect when classifying programs is the intended beneficiaries. PMI (2013b) mentions that certain benefits are realized by the organization delivering the program benefits, whereas others are delivered to external customers.

If we partially align ourselves with the PMI understanding of programs and ignore the portfolio type of programs, we can classify programs based on five key factors:

1. Program motivation, that is, whether the program came into existence from existing projects, based on the realization that they are linked together through strategic goals and grouping them will increase chances of goal accomplishment, *or* was the new initiative based on a strategic decision with no prior footprints.
2. Benefits delivery mechanisms, that is, whether the program is delivering incremental benefits *or* will deliver benefits at the end of its life cycle, when all component projects will close.

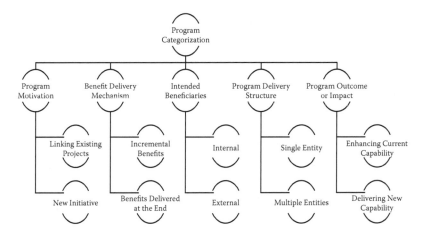

FIGURE 6.2
Program categorization.

3. Intended beneficiaries, that is, whether the customers are internal to the organization *or* are external to the delivering entities.
4. Program delivery structure, that is, the program is delivered by single entity *or* has multiple entities involved.
5. Program outcome or impact on organization, that is, whether the program is enhancing a current capability for the beneficiary *or* delivering a new capability.

These typologies as depicted in Figure 6.2, have to be well understood to ensure that the most effective management and governance practices are applied so that a successful program delivery is guaranteed. Pellegrinelli (1997) mentions that "discussions with programme managers and senior managers within organizations and analysis of their practices is that the rationales for, and benefits expected from, programmes lead to different programme management structures" (p. 143).

PROGRAM MANAGEMENT

Where different projects are contributing to the same objectives it is more useful to group them under one program to manage them more effectively in order to deliver the required benefits. In such circumstances the program setup defines the governance regime for the individual projects, and the program manager acts as the sponsor of the project, and performs the defined governance role (Müller 2011). Programs require efforts by management to coordinate and control the projects and their outcomes as well as managing the overall benefits that the programs have to deliver.

PMI (2013b) mentions program management as the "application of knowledge, skills, tools, and techniques to a program to meet the program requirements and to obtain benefits and control not available by managing projects individually. It involves aligning multiple components to achieve the program goals and allows for optimized or integrated cost, schedule, and effort" (p. 6). The program management focus as per PMI (2013b) is on "strategic benefits, coordinated planning, complex interdependencies, deliverable integration, and optimized pacing" (p. 7).

APM (2009) relates to program management by mentioning that "Program management is the co-ordinated management of related projects, which may include related business-as-usual activities that together achieve a beneficial change of a strategic nature for an organisation. What constitutes a program will vary across industries and business sectors but there are core program management processes" (p. 6).

OGC (2009) defined program management as the "action of carrying out the coordination, direction and implementation of a dossier of projects and transformation activities (i.e., the program) to achieve outcomes and realize benefits of strategic importance to the business" (p. 6).

The core of all three definitions is the coordinated management of related projects resulting in the achievement of strategic goals and benefits, expected from the program, for the organization. Portfolios and programs act as entities to oversee the activities being carried out at the project level. The difference between a program and a portfolio is that the projects in the program are linked together through a shared objective or benefit and are interdependent, whereas the portfolio component may not necessarily be interdependent or directly related (PMI 2013a).

Programs do not manage projects, which is a project management responsibility; however, they ensure that the projects are aligned with the strategic objectives of the organization and that any issue or risk that cannot be managed at the project level is resolved at a higher level.

Program management ensures that the collection of projects, interdependent or not, should provide the required results expected from them as a group. This does not necessarily mean that the program is not concerned with the individual project; however, the main focus is to resolve collective issues and deliver collective benefits. Programs also have the responsibility to oversee that the projects are being managed according to the organizational standards and is progressing as per the committed plan.

PROGRAM LIFE CYCLE

The concept of temporariness in programs, having a predefined start point and end point, necessitates the design of a life cycle. Temporary organizations have a beginning, middle, and end, and the idea of birth to death is the core difference between permanent and temporary organizations (Lundin and Söderholm 1995). Also, it is logical to presume that the program outcomes cannot be generated in one step. The outcome of the program needs to build, through component deliveries, and a conscious and planned effort will be required to deliver the program objectives. This stagewise program approach to move from definition to benefits delivery and closure is termed the *program life cycle.*

Pellegrinelli et al. (2007) state that "programmes are conceived as frameworks or structures, and so temporal or with indeterminate time horizons, rather than having linear life-cycles akin to projects" (p. 42). Thus, program life cycles are significantly different than project life cycles. The close linkage between programs and organizational strategy dictates that the program life cycle supports the evolving organizational context. In addition, there are characteristics such as a high level of uncertainty, a need for continuous integration, and a focus on benefits delivery, which has a strong influence on the structuring of the program life cycle. This influence directs that the program life cycle should have the following three characteristics (Pellegrinelli 1997; Thiry 2007):

1. The program life cycle should have certain processes such as cyclic or iterative, where there is a need for evaluation/feedback from stakeholders and integration with the organizational vision and direction.
2. The program life cycle should also have an internal focus where emphasis should be on collaboration and integration between program components.
3. The overall program life cycle should allow for revisiting different phases as dictated by the program context.

FIGURE 6.3
A program life cycle.

While aligning with research (Pellegrinelli 1997; Thiry 2007; PMI 2013b) we can divide program life cycle into the following four phases:

1. Formation
2. Organization
3. Benefits delivery
4. Closure

These phases (Figure 6.3) are iterative in nature and might overlap and be revisited during the life cycle.

Formation

The starting point of this phase is the decision to initiate a program to fulfill a strategic objective. This decision is generally devised, reviewed, and approved at an executive or portfolio level, based on the program's feasibility.

The program sponsor and manager are assigned at this stage, as the phase sets up the foundation for further program activities. Based on the information available, certain activities such as initial resource planning, cost, and scope estimates are conducted at this stage. Another objective of this phase is to formally define the program road map and approve the program charter. The deliverables of this phase are utilized in the following phases, where more detailed plans are created.

The activities carried out in this phase are iterative in nature as an outcome of one activity may result in review and revision of another output.

This phase overlaps with the organization phase, which is generally initiated after the program charter is approved.

Organization

This phase is iteratively executed with the formation stage and utilizes the deliverables, such as a program charter of the formation stage. Some of the core planning and structuring activities carried out in this phase include:

1. Definition of the program's management structure.
2. Development of the program management plan, which includes all subsidiary plans.
3. Deployment of the initial program team.
4. Establishment of the program's formal governance structure.
5. Definition of operational procedures and standards.
6. Finalization of key performance indicators (KPIs) and critical success factors (CSFs).
7. Identification of the program's component projects.
8. Identification of roles and responsibilities.

The program management plan is formally approved by the governance board. The program moves to the benefits delivery phase after this approval. As the program unfolds during the subsequent stages, the deliverables of this phase are revised, through formal processes, in order to stay aligned with the objective to deliver the chartered program objectives.

Benefits Delivery

This phase of the program focuses on, as the name suggests, delivering program benefits. Projects are initiated, executed, and closed, and the resulting outcomes are integrated to deliver benefits to the beneficiary organizations. Depending on the benefits delivery model, the project outcomes will be integrated into deliverable(s) and transitioned to the beneficiaries, iteratively or at the end of this phase, where the benefit(s) will be realized and sustained.

Another core activity of this phase is consistent alignment with the organizational strategy. This alignment is ensured during the overall program life cycle, as a shift in the organization's direction will result in the revision of deliverables and plans that were devised in the prior stages.

Certain component projects might be terminated, and new projects might be added to the program. The program's management and governance structure should be flexible and adjusted to the realigned program direction and plans.

This phase is cyclical in nature, as the component projects are reviewed for performance and delivered benefits are consistently assessed, revised, and transitioned to beneficiaries. In addition, the plans developed in the earlier phases are revised in order to be aligned with the existing program context.

Closure

Once all expected benefits are delivered, or the need for the program ceases to exist, the program is closed. Postprogram appraisals are conducted by the governance board at this stage to review the overall benefits delivered by the program. These appraisals assess the level of program success based on KPIs defined during the prior stages. This activity also assists in documenting the lessons learned that are beneficial for future similar initiatives.

Based on the organization's strategic direction, the program's incomplete work or projects might be transferred or transitioned to another program, which will once again go through a similar life cycle.

The program is officially closed at this stage, and the resources are reassigned to other activities and entities, for example, the sustaining organization responsible that will oversee the benefits delivered by the program.

EXTERNAL CONTEXT OF THE PROGRAM

Programs are created to achieve strategic organizational objectives that cannot be achieved through routine operations. Program characteristics, such as the focus on benefits delivery, the need for integration, acting as an agent for change, a strategic focus, and a high level of uncertainty, makes it sensitive to its context.

PMI (2013b) proposes to conduct environmental assessments, and mentions that there "are often influences inside and outside of the program that have a significant impact on the program's ultimate success. Some of the influences from outside the program are internal to the larger organization, and some come from completely external sources. Program managers identify these influences and take them into account when developing and managing the execution of the program in order to ensure ongoing stakeholder alignment, the continual alignment with organizational goals, and overall program success" (p. 30).

Pellegrinelli et al. (2007) defined the program context as "a dynamic cultural, political, and business environment in which the program operates" (p. 41). They further mention that "Programmes shape and coordinate projects and related activity in pursuit of organizational goals and benefits in the context of a dynamic organisational environment" (p. 52). Thus, the understanding of program context is imperative.

The program's context can be reviewed for two perspectives:

1. Internal context, that is, the attributes of the task itself.
2. External context, that is, the attributes of the environment within which the task is executed.

The internal context is important to understand because the core attributes of the program define the type of management and governance structure that is required to effectively deliver the program benefits. This context has already been discussed in this chapter when we reviewed program characteristics, program typology, and program life cycle.

In addition to the internal context, the external context has a strong influence on the program's direction as well. PMI (2013b) lists certain environmental factors that have an influence on the program and its governance, as well as the management framework.

Table 6.1 lists certain factors, external to the program, which can have a significant impact on the program, its management and governance structure, and its ultimate success. Some of these factors have a direct influence on the program; however, others might have an impact on the organization, which might in turn have an impact on the program.

Shao (2010) conducted extensive research, and proposed four environmental factors that define the program's external context:

1. Organizational fit
2. Program flexibility
3. Organizational stability
4. Resource availability

TABLE 6.1

Program's External Context

Program's External Context	Internal/External
Organizational strategy	Internal to organization; can be influenced by external factors
Organizational culture	Internal to organization
Organizational stability	Internal to organization; can be influenced by external factors
Resource availability	Internal to organization
Legislative and regulatory requirements	External to organization
Cultural and geographical influences	External to organization
Economy and market conditions	External to organization
Wider range of stakeholders	Internal and external to organization

Organizational Fit

This factor relates to the alignment between the program and its organizational context. Organizational culture, power lines, strategy, stakeholders, business direction, and so on, all have an influence on the program. Shao (2010) states that the alignment between program and the organizational context increases the chances of program success.

It is important for the program and the supporting entities to understand the power lines and remain aligned with the organizational strategy. In addition, the program culture has to relate to the overall organizational culture, for example, if the organizational culture promotes and rewards innovation then a similar mechanism needs to be put in place within the program boundaries.

Program Flexibility

Program flexibility can be defined as the adaptability of the program's governance and management structure to the external environment. Rigid management and governance mechanisms cannot respond to evolving program context.

The flexibility of the program can be measured through:

1. Flexibility of management and governance structure
2. Flexibility of procedures and processes

Organizational Stability

Programs exist within the organizational umbrella, thus the stability of the organization has a strong influence on the program. This factor determines the stability of the organization in terms of organizational structure and program-related processes. An unstable organization always possesses a threat to the ultimate success of the program.

The stability of the organization can be measured through:

1. Stability of the parent organization structure
2. Stability of the program-related process in the parent organization

Resource Availability

This factor represents the support available from the organization in terms of resources. The availability of resources when required by the program has a strong influence on the program's execution and delivery mechanism. A program that is short in terms of resources will always be in a critical situation in terms of benefits delivery. Resources include but are not limited to human resources, financial resources, space, location, and other facilities.

Resource availability includes:

1. Human and material resource availability
2. Funding availability

SUMMARY

Programs can be considered as temporary organizations that are initiated to meet strategic objectives and delivery benefits. The mechanism of delivery depends on the scope and objectives of the program. However, all programs follow a life cycle that starts with the program's formation and organization, before delivering the objectives, and resulting benefits, through its components. Figure 6.4 provides a comprehensive overview of a program and its context.

Because of the complex nature of program undertaking, different phases of the life cycle overlap and are revisited as the program is elaborated and executed. The program closes when all the objectives expected

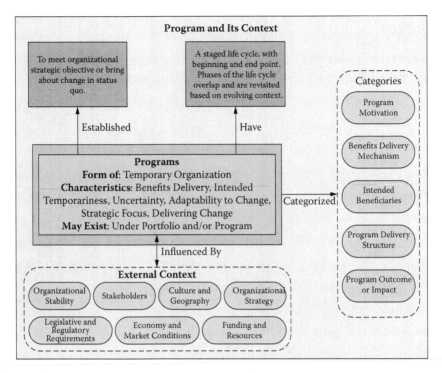

FIGURE 6.4
A program and its context.

of the program are met or the need for the program ceases to exist. In some cases, the program deliverables are utilized at the end of the program life cycle to realize the program benefits. However, there are other cases where the program's benefits are realized years after the program has delivered its objectives.

Programs have certain unique characteristics and are categorized based on different attributes. The understanding of a program's characteristics and its internal context is important, so that management and organizing practices aligned with the specific needs of the program, under consideration, can be adapted. This alignment increases the chances of successful delivery of program objectives and resulting benefits.

There are various external factors that have an impact on the program. One example of such impact is the North-South Metro Line project, one of the biggest infrastructure projects in the Netherlands, which is being built in Amsterdam and is planned for commissioning in 2017 (Staal-Ong and Westerveld 2010). The project team had to plan for the livability and safety of the people, impact on environment, and accessibility of buildings in the city. Different measures were taken to ensure that stakeholders remain satisfied with the project and its execution. One of the measures was to develop a livability plan, which defined how people can apply for a substitute workplace, alternate housing, stay in hotels, and compensative measures for people living in the vicinity. The influence of stakeholders and other such factors, which have been discussed in this chapter, have to be thoroughly considered during the execution of programs.

REFERENCES

Artto, K., Martinsuo, M., Gemünden, H. G., and Murtoaro, J. (2009). Foundations of program management: A bibliometric view. *International Journal of Project Management,* 27, 1–18.

Association for Project Management (APM). (2009). *APM Body of Knowledge* (Fifth Edition). Buckinghamshire, UK.

Gray, R. J. (1997). Alternative approaches to programme management. *International Journal of Project Management, 15*(1), 5–9.

Global Alliance for Project Performance Standards (GAPPS). (2011) *A Framework for Performance Based Competency Standards for Program Managers.* Sydney, Australia: Global Alliance for Project Performance Standards.

Lundin, R. A. and Söderholm, A. (1995). A theory of the temporary organization. *Scandinavian Journal of Management, 11*(4), 18.

Müller, R. (2011). Project governance. In Pinto, J., Morris, P., and Söderlund, J. (Ed.), *Oxford Handbook of Project Management.* Oxford, UK: Oxford University Press.

Murray-Webster, R. and Thiry, M. (2000). Managing programmes of projects. In Turner, J. R. and Simister, S. J., (Ed.), *Gower Handbook of Project Management* (Third Edition). Aldershot: Gower.

Office of Government Commerce (OGC). (2009). *Managing Successful Projects with PRINCE2.* London, United Kingdom: The Stationery Office (TSO).

Pellegrinelli S. (1997). Programme management: Organising project-based change. *International Journal of Project Management, 15*(3), 141–149.

Pellegrinelli, S., Partington, D., Hemingway, C., Mohdzain, Z., and Shah, M. (2007). The importance of context in programme management: An empirical review of programme practices. *International Journal of Project Management, 25,* 41–55.

Project Management Institute (PMI). (2013a). *A Guide to the Project Management Body of Knowledge* (Fifth Edition). Newtown Square, PA: Project Management Institute.

Project Management Institute (PMI). (2013b). *The Standard for Program Management* (Third Edition). Newtown Square, PA: Project Management Institute.

Shao, J. (2010). *The impact of program manager's leadership competences on program success and its moderation through program context.* Thesis for Doctor of Philosophy in Strategy, Program and Project Management, Lille, France: SKEMA Business School.

Staal-Ong, P. L. and Westerveld, E. (2010). The North-South Metro Line, managing in crowded historic Amsterdam. In Turner, J. R., Huemann, M., Anbari, F. T., and Bredillet, C. N. (Ed.), *Perspectives on Projects.* New York: Routledge.

Thiry, M. (2007). Program management: A strategic decision management process. In Morris, P. and Pinto, J. K. (Ed.) *The Wiley Guide to Project, Program, and Portfolio Management.* John Wiley & Sons.

Turner, J. R. and Müller, R. (2003). On the nature of project as a temporary organization. *International Journal of Project Management, 21*(1), 1–8.

Vereecke, A., Pandelaere, E., Deschoolmeester, D., and Stevens, M. (2003). A classification of development programmes and its consequences for programme management. *International Journal of Operations & Production Management, 23*(10), 1279–1290.

Section III

Program Governance

7

Program Governance

PMI (2013) defines an executive sponsor as a senior executive who is responsible for the success of an authorized program activity. The executive sponsor is a program governance role; however, program governance, as a subject, is a relatively emergent concept. This book, which you have in your hands, can be considered as one of the first books on this subject. This contradiction that exists between the importance of the governance role as a key factor for program success and its lack of visibility in literature is incomprehensible. It could be due to the literature on program governance being hidden under other themes such as portfolio management, steering committees, and governance boards; however, there is limited literature available on the subject of program governance. Thus, it is important to define this term, so that the foundation of this concept can be established for the coming sections.

DEFINING PROGRAM GOVERNANCE

Program governance is considered as a framework that oversees program activities, and ensures the integration and alignment of constituent projects. The program governors act as an escalation point for the program's management team and address issues and risks that cannot be handled at the management level. This framework exists within the corporate governance framework of the organization and can be considered an integral part of it.

Ward defines program governance as a "process or set of processes used by an organization to define, develop, manage, monitor and close-out a program. It is primarily aimed at ensuring that the program's objectives are met and benefits delivered, and it includes a process for terminating a program if it appears the program will not meet identified objectives and benefits" (Ward 2008, p. 338).

From a more practitioner perspective, Hanford defined program governance as "a combination of individuals filling executive and management roles, program oversight functions organized into structures, and policies that define management principles and decision making" (Hanford 2005).

Reiss et al. (2006) state that "programme governance consists of the leadership and organisational structures and processes to ensure that the programme sustains and extends the organisational strategies and objectives."

If we look at the definitions above we can extract the two elements that form the core of program governance:

1. A set of policies, procedures, and guidelines aimed at efficient benefits delivery.
2. The structure or entity that oversees that program.

There has been some significant work carried out by the Project Management Institute (PMI) related to program governance. PMI (2013) mentions that program governance, which fits within the overall framework under which programs are managed, provides the program team with the management framework that constitutes process, policies, structure, and decision-making models. PMI (2013) defines program governance as a framework that "covers the systems and methods by which a program and its strategy are defined, authorized, monitored, and supported by its sponsoring organization" (p. 51). Program governors provide support to the program, and monitor and control the progress of the program throughout its life cycle.

Program governance ensures that the decision-making and benefits delivery activities are focused on achieving program goals in a consistent manner, while keeping the program aligned with the organizational strategy. It also ensures that program risks are addressed appropriately, and that the program stakeholders are satisfied with the program's progress.

It is important to realize that program governance must consider the broader context of enterprise governance and recognize how to report and collaborate with the enterprise governance function. Program governance is linked with other governance areas and assets, thus it cannot be considered as an isolated function.

Looking at the definitions discussed above, we have a more comprehensive definition of program governance that is relevant to our discussion.

Program governance is an oversight function that encompasses the program life cycle, and provides the program team with a structure, processes, decision-making model, and tools for managing the program, while keeping the program aligned with the organizational strategy. The governing entity monitors, supports, and controls the program in order to ensure efficient benefits delivery.

WHY SHOULD PROGRAMS BE GOVERNED?

I acted as a consultant to organizations in the Middle East for almost a decade. The model that most government organizations adapt there is based on outsourcing projects and programs to vendors for delivery. The vendor takes on the responsibility of project/program management and the actual work of the project/program in order to deliver the defined objectives/benefits. One key role that certain organizations ignore is the role of governance. Organizations assume that the vendor is responsible for delivery, and if the objectives are not delivered, the penalty clauses in the contract will come into action. This assumption, even though true, does not guarantee the success of the initiative.

If these organizations have proper governance frameworks in place, they would not be waiting for the vendor to fail; rather, they would proactively monitor, control, and support the initiative and resolve the issues before they arise. A proactive approach would increase the probability of the success of the initiative.

The governance of a program follows a similar pattern to that of corporate governance; however, the focus can be a bit different because of the temporary nature of these initiatives. Program governance starts with a definition of the objectives for the program and encompasses support and control needed to meet those objectives throughout the program life cycle.

The way a program proceeds and succeeds is not entirely in the control of the program manager and program team. This situation is because there are various influences, which are outside the program manager's domain, and they need intervention from the executive management. There might be additional influences, such as changes in organizational strategy, which have to be communicated to the program, so that the program remains aligned to these changes. This is the reason why program managers need the program governor's directions, support, and control to make the program successful.

Programs require a management and decision-making framework, which can assist the program team to administer and manage the program during the life cycle. The development and implementation of this framework is another facet that the program governors address. Program governance provides the program management team, the necessary structure, processes, tools, and techniques, as well as the decision-making model, which are required for the program to execute and deliver the required benefits to the organization. This framework assists the program management team to work in a standardized manner, while ensuring that decisions are made at the appropriate level of authority without creating cumbersome bureaucracy.

Another reason for setting up governance frameworks is due to the supposition that organizations execute multiple initiatives at the same time. These initiatives consume time and resources of the organizations and are a means of achieving strategic organizational goals. Thus, they should be governed by some sort of an organizational entity, which ensures that interinitiative constraints are resolved. Each initiative gets the required support, and it is properly monitored and controlled (especially when things go wrong) by the program governors during its life cycle in order to increase its chances of success.

PMI (2013) states that effective program governance supports the success of a program by:

1. Ensuring the strategic alignment of the program with organizational goals.
2. Providing a management framework that decreases the uncertainty related to the program road map and benefits delivery.
3. Managing the stakeholder expectations, while keeping them abreast about the program progress.
4. Providing processes, tools, and guidelines to be utilized by the program team for consistent delivery of program outputs.
5. Reducing the ambiguity around decision making through provisioning of a decision-making model.
6. Providing clear roles, responsibilities, and an accountability framework that ensures that each program team member understands his or her duties in program organization.

7. Providing a clear mechanism to initiate, execute, and close program components, while supporting the program components during their life cycle.

8. Creating a centralized authority that can assist in the management of escalated risks and issues that cannot be handled at the management level.

9. Creating a consistent program audit and review mechanism that enables the organization to review program progress, its current viability, and the type of resources required to meet the program objectives. Such audits also ensure the program conformance with organizational standards.

Programs are complex and highly uncertain undertakings and require a governance mechanism that can assist them in responding and adapting to changing context as the program progresses. Without this senior management oversight, there is a high probability that the program might misalign with the evolving strategic objectives resulting in significant damages to the organization and eventual termination of the program.

Program governance processes, through a continuous review mechanism, enables the organization to monitor and control the program activities while providing the necessary support from different aspects. PMI (2013) states that program governance "provides an important means by which programs seek authorization and support for dynamically changing program strategies or plans in response to emergent outcomes" (p. 52). Figure 7.1 shows different dimensions of program governance that have been discussed in this section.

PROGRAM GOVERNANCE PERSPECTIVES

To recognize the benefits that program governance provides, we need to realize the different perspectives that program governance incorporates to increase the chances of a program's benefits delivery.

Based on the industry, as well as research experience, there are four main perspectives as depicted in Figure 7.2, related to program governance. Each of the following perspectives provides an insight into different motivations of organizations when implementing a program governance framework:

FIGURE 7.1
Program governance dimensions.

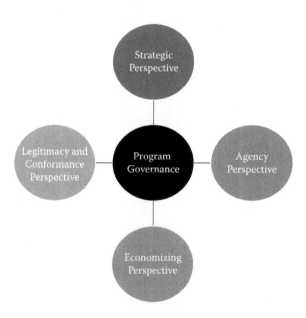

FIGURE 7.2
Perspectives of program governance.

1. Economizing perspective, which has its roots in transaction cost economics (TCE) theory, where programs are considered as transactions.
2. Principle–agent perspective, which has its foundations in agency theory, where programs are considered as an agent.
3. Legitimacy or conformance perspective, where the focus is on ensuring that legitimate work is carried out within the program and by the program.
4. Strategic perspective, where the focus is on strategic alignment and benefits delivery.

From a TCE perspective, programs are viewed as transactions, and the governance goal in this case is to economize on transaction costs. The core reason for forming programs is to ensure that the coordinated management of a program's components results in achieving benefits that cannot be achieved if they are managed separately. This can be achieved through the efficient sharing of resources, minimizing the costs of contracting, and the optimization of internal processes and methods.

From an agency perspective, programs are viewed as agencies where the program manager acts as an agent for the principal, which can be any entity authorizing the program manager to act on their behalf. The focus in this case is to monitor and control the actions of the program's management team through incentives, interest alignment, and reducing the knowledge gap. The objective of governance in these cases is to minimize agency costs by addressing the problems of moral hazards and adverse selection. This might have been the perspective of my customer when he insisted on implementing a governance framework in order to penalize the vendors.

Legitimacy or conformance perspective addresses the concern that the program and its activities should be legitimate and comply with the rules and regulations defined by the law. The activities of the program should also be within the ethical boundaries defined by society. Within a corporate governance framework, conformance to external requirements is considered one of the most important functions. The legitimacy perspective is the focus of all Information Systems Audit and Control Association (ISACA) references, which have a major focus toward the control dimension of governance (ITGI 2003; ITGI 2007; ISACA 2010).

The strategic perspective ensures that the program is in constant alignment with the organizational strategy and is delivering the expected benefits. The focus here is on ensuring that the right projects are selected, and if a project is not contributing to the program(s) benefits, it is terminated. Any shift in the organizational strategy, or context, is reflected in the program's implementation strategy. The perspective has a strong alignment with the portfolio management practices.

An effective governance framework can be designed when all the above perspectives are considered together, with varying degrees of significance. This ensures that an economized governance model is created, which addresses the internal problems related to program governance while focusing on the external requirements and factors.

SUMMARY

Organizations tend to implement governance based on different motivations, while employing various approaches. Some organizations focus on governance using behavioral control (Müller 2010) through the implementation and governance of processes. There are other organizations that focus on governing the environment by employing different mechanisms, roles, and institutions of governance (Turner and Keegan 2001; Müller 2011). Based on my experience, both these perspectives are important to the design of an effective governance framework within an organization.

Program governance exists within the corporate governance framework. However, the program's governance framework has to be relatively flexible, and should take into account various factors. These factors include changing program context or different attributes of the program. This flexibility ensures alignment between the program's context and its governance requirements. This alignment, in turn, increases the probability of the successful delivery of program objectives and related benefits.

During the initial phases of the program life cycle, the focus of governance is more toward providing the processes, tools, and decision-making framework. However, as the program progresses, the focus shifts toward monitoring, supporting, and controlling the program through different governance functions and required mechanisms. This shift in focus of governance based on the program's life cycle is depicted in Figure 7.3.

While designing a program governance framework a few questions must be answered:

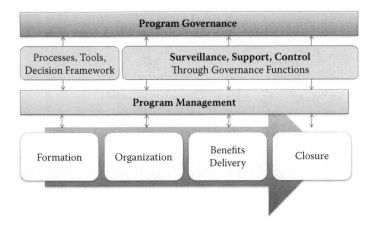

FIGURE 7.3
Program governance.

1. What is to be governed or what aspects of the program should be governed, that is, what are the domains of governance?
2. What functions will the program governance mechanism fulfill?
3. What will be the governance structure, that is, what institutions or roles will be governing the program?
4. What will be the mechanism of governance, that is, how will the governance function be employed?
5. What are the factors that will affect the governance mechanism?

Answers to these questions will result in the formation of a program governance regime that fulfills the organizational needs of: (a) strategic alignment, (b) reduction in the agency costs, (c) minimizing of transaction cost, and (d) ensuring the legitimacy of the transaction or agency under consideration.

I will focus on each of the above-mentioned areas in the following chapters, so that you, as a practitioner or researcher, can recognize the different lenses through which governance can be understood in its totality.

REFERENCES

Hanford, M. F. (2005). *Defining Program Governance and Structure*. IBM. Retreived March 13, 2013 from http://www.ibm.com/developerworks/rational/library/apr05/hanford/.

Information Systems Audit and Control Association (ISACA). (2010). *CGEIT Review Manual 2010* (First Edition). Illinois: ISACA.

IT Governance Institute (ITGI). (2003). *Board Briefing on IT Governance.* Illinois: IT Governance Institute.

IT Governance Institute (ITGI). (2007). *COBIT 4.1.* Illinois: IT Governance Institute.

Müller, R. (2010). Project governance. [Monthly Column—Series on Advances in Project Management]. *PM World Today XII*(III), 1–6.

Müller, R. (2011). Project governance. In Pinto, J., Morris, P., and Söderlund, J. (Ed.), *Oxford Handbook of Project Management.* Oxford, UK: Oxford University Press.

Project Management Institute (PMI). (2013). *The Standard for Program Management* (Third Edition). Newtown Square, PA: Project Management Institute.

Reiss, G., Anthony, M., Chapman, J., Leigh, G., Payne A., and Rayner, P. (2006). *Gower Handbook of Programme Management.* Aldershot: Gower.

Turner, J. R. and Keegan, A. (2001). Mechanisms of governance in the project-based organization: Roles of the broker and steward. *European Management Journal, 19*(3), 254–267.

Ward, J. L. (2008). *Dictionary of Project Management Terms* (Third Edition). Arlington, VA: ESI International.

8

Program Governance Domains

I initiated my first technology start-up in 2001. The company did not deliver as I had initially planned. However, the lessons learned from this experience really helped me in my future initiatives. One item that I clearly missed was dividing my attention on different organizational verticals based on the changing market and organizational context. I thought a quality product would sell itself; thus, my attention was totally focused on developing products with high quality.

We had 10 technical resources that were developing solutions; however, we only hired one individual (who had limited sales experience) to sell our solutions. Here I had a quality product with literally no one to sell it. You can see what I missed. I wish someone had guided me in realizing that there were different domains I had to focus on while running the company, or had I recognized it myself I might have done a bit better. I do not want something comparable to happen to my program governors.

In all of my consulting assignments, one of the key aspects that I try to emphasize is to make an organization recognize the areas that should be governed. Certain responsibilities have to be transferred to the program's management team. Others might be shared between the governance and management teams based on predefined thresholds. In essence, the domains covered in the coming sections fall within the scope of program governance with varying degrees of focus. The significance of each domain depends on various factors, such as the life cycle stage of the program or the overall program's performance.

It is important to realize that these domains are not mutually exclusive when it comes to implementation. There will always be some level of overlapping when actual mechanisms of governance are implemented. The distinction provided in this book is for the purpose of clarity, for pracademics, to determine the areas that program governance has to focus on explicitly.

PROGRAM STRUCTURE AND PROCESSES

One of the main domains of governance is to create an environment under which the program is governed, executed, and managed smoothly. There are two facets related to this element. One is to create a program management and governance structure that includes defining the roles, responsibilities, and reporting structures, whereas the other is concerned with setting up the processes, policies, decision-making models, escalation mechanisms, templates, and tools for program management. All this information is included in the program governance plan.

Having a clear understanding of "how things should work" and "how the program is structured" simplifies the task of the program management team, as they will have a relatively clear idea as to how the program should be smoothly routed, and how the decisions will be made during the course of the program. This understanding also reduces the potential for blurred boundaries between different entities, as each entity will have a clear idea about its responsibilities, authority, and domain of influence.

Processes play a vital role in streamlining business operations and relationships between different organizational entities. Thus, attention must be paid to the process part of the program organization. Assuring that the program team is following the defined processes and that the program governance mechanism is effectively implemented is a responsibility of the program governance team. The Norwegian QA framework, during pre-initiation activities, focuses on ensuring that the team is following the defined mechanism through reviews of standard documents at QA 1 and QA 2 stages (Klakegg et al. 2008). QA 1 and QA 2 are the two stage gates when the quality of the deliverables is accessed. As a control dimension, governance should evaluate the proper implementation of program management processes and tools.

PROGRAM RESOURCES

The management of program resources is the responsibility of the program management team; however, governance oversight should be provided from the program governors. Turner and Keegan (2001) refer to the role of the steward, who has the responsibility of the optimal assignment of resources to the team. The provision of resources for the program team, in order to deliver program benefits, is a governance responsibility. Thus, the program governance team has to make sure that resources are adequately available for the program and its constituent projects.

In addition, the program governance team has to ensure that the resources are being utilized in an optimal manner based on the defined plan. The program's financial expenditures have to be reviewed periodically to ensure that the program spending remains aligned with the financial plan. It also helps in unearthing any unwarranted activities that might create significant issues in the future. In terms of program spending, any significant deviations, above the threshold level of the plan, have to be reported to, reviewed, and approved by the governance team.

PROGRAM STRATEGY

Ensuring the strategic direction of the program comes within the scope of program governance. The program manager should have a strategic perspective while managing the program. However, his or her horizon is concentrated on the program's benefits, accordingly he or she strategizes—based on the same vision.

The program governance team has a broader organizational perspective, and hence they can ensure that the routine activities and their resulting program deliverables are in alignment with the organizational strategy. In addition to this, the program governance team iteratively links the organizational strategy with the program strategy.

This linkage can be in both directions. The organizational strategy will have an impact on the program's direction; however, any significant events in the program can influence the organizational strategy as well. For example, successful closure of major components that delivered a strategic solution, before the defined deadline, might influence the organization to go to the market before the planned launch date.

PROGRAM DECISIONS

The program manager and team look toward the program governors to provide guidance and support regarding certain program decisions. The program governors also evaluate the decisions that are made by the program team over certain aspects. Decisions are generally evaluated through outcomes or through the decision-making process.

The program governance team conducts the following activities related to program decisions:

1. Define the framework for program decisions by the program team.
2. Provide a mechanism through which decisions made by the program team are reviewed.
3. Define the domains that come under the direct influence of the program governance team.

The Office of Government Commerce (OGC) gate review 0 assesses the decisions made by the program team related to the program as well as projects within the program (OGC 2007). The Project Management Institute (PMI) also recommends program and project managers to document the decisions that are made as the governance function is expected to audit the program and related projects, and the decisions will be evaluated as part of the audit process (PMI 2013a).

ESCALATED RISKS AND ISSUES

Management of risks and issues is a program management level responsibility. However, there are certain risks and issues which are outside the direct control and scope of the program team that have to be escalated to the program governance team. Resolution of these risks and issues are under the scope of the program governance team (Stretton 2010).

The escalation can be for actions from the governance team or for information only. In an ideal world, the escalation communication can be agreed to upfront. However, realistically speaking, the program team has to make some judgmental calls. In some cases, the program management team will have to see whether all necessary actions have been taken to resolve the issues before escalation. There might also be situations where they may have to evaluate whether resolution of a particular issue is beyond the level of the management team. In the case of high impact issues, even if the management team is resolving the issue at their level, they still escalate the issue in order to bring the situation to the knowledge of the governance team.

The program needs to have a defined framework for escalation, as leaving it only to the judgment of the program team can result in redundant activities at the governance level. A framework ensures that issues and risks are not escalated either too early, resulting in unnecessary activities at the governance level, or too late, resulting in a disastrous situation or missed opportunity. I am calling it a *framework*, not a *process*, because the framework will provide guidelines and principles for escalation while allowing some level of subjectivity at the individual level. Numbers tell everything, but it is the individual who interprets and responds.

Following are some examples of risks and issues that may require escalation:

1. Conflicts related to resource assignment on components.
2. Critical issues related to key projects that need immediate intervention.
3. Resolutions that are outside the tolerance level, for example, will cost or will take longer.
4. Issues that might have an impact on another component.
5. Need for prioritization on the component level.
6. Influences from external stakeholders, such as other departments, organizations, or government entities.

Escalation means to involve the governance team to gather their support for issue resolution or just "keeping them posted" about the situation. This will ensure that the program's management and governance team continue working in a collaborative and synergetic environment with no surprises.

PROGRAM PROGRESS

The overall progress of the program should be consistently governed by the program governors. The executive sponsor should scrutinize the program, provide support to the program team, or control the progress when the program shows signs of deviation.

The progress of the program can be measured from the following perspectives:

1. Program schedule
2. Program financial spending

3. Program resource utilization and requirements
4. Program's consistent strategic alignment
5. Program benefits delivery and realization status
6. Progress of component projects
7. Interaction between interdependent projects
8. Issues and overall risk exposure of the program

Based on the status review of the program, as well as other factors, the program governors may decide to continue or to terminate the program or one of its component projects. Monitoring the program status helps align the program, and its projects, with the organizational context as well. Stretton (2010) considers periodic program reviews of cost and progress as an important governance function.

PROGRAM BENEFITS

The reason for initiating a program is to achieve organizational strategic objectives; thus, this aspect of the program should be at the center of attention of the program governors. Defining and reviewing the program objectives are considered to be an important function of governance.

The program manager has the responsibility to ensure the realization of program benefits. However, the governance team supports the program manager by providing guidance in terms of:

1. Identification of program benefits.
2. Planning for effective benefits realization.
3. Aligning the program benefits with the organizational strategy on a consistent basis.
4. Reviewing the quality of the deliverables produced by the program and its components that will contribute to benefits realization.
5. Assessment of the impact of requested changes on the program's expected benefits and associated plans.
6. Resolve interdependencies between the components and their outcomes to ensure effective benefits delivery.
7. Monitor, support, and control the program from a benefits delivery and realization perspective.

8. Liaison with the beneficiaries in terms of preparing them for sustaining the delivered benefits.

All of OGC's gate reviews focus on the strategic objectives that the project or the overarching program is focused toward achieving (OGC 2007). PMI refers to the management of program benefits, which are achieved through the collective objective accomplishment of projects, as a program governance responsibility (PMI 2013a).

PROGRAM STAKEHOLDERS

While identifying the role of broker, Turner and Keegan (2001) mention that one of the major assets that has to be governed are the program stakeholders, such as the program. PMI (2013b) states that stakeholders "represent all those who will interact with the program as well as those who will be affected by the implementation of the program" (p. 45). Program stakeholders can be internal or external to the program or can be directly or indirectly impacted by the program outcomes.

The program acts as a vehicle or driver for change. The impact that the program creates is generally large and can directly or indirectly affect a wide range of stakeholders. Even though stakeholder engagement is a program management role, there are certain stakeholders with whom managing their expectations is beyond the direct control of the management team. In such cases, the management team will require assistance and guidance from the governance entities. The expectations of these stakeholders have to be managed, or in some cases influenced, by the program governance team. While relating ourselves with PMI (2013b), these stakeholders can be the:

1. *Funding organization*—The part of the organization or the external organization providing funding for the program.
2. *Performing organization*—The group that is performing the work of the program through component projects and nonproject work.
3. *Beneficiaries*—The individual or organization that will use the new capabilities/results of the program and derive the anticipated benefits.

4. *Suppliers*—Providing products and services to the program, who can be affected by changing policies and procedures of the program, as well as changes in program strategy.

5. *Governmental regulatory agencies*—Providing the regulations and legal framework within which organizations and programs function.

6. *Nongovernmental standardization authorities*—Which set the standards and requirements for adherence.

7. *Competitors*—Who keep a strong eye on the program outcomes as it might have an impact on their strategy.

8. *Impacted entities*—Those who perceive that they will either benefit from or be disadvantaged by the program's activities.

9. *Other groups*—Those representing consumers, environmental agencies, and other entities that have an interest in the program.

Association for Project Management (APM) principles refer to proper management of, and reporting to, stakeholders based on their importance to the initiative (APM 2007). The management of organizational stakeholders, who can influence the program condition and outcomes, has to be carried out by the executive sponsor. To generate sustainable business and growth, clients and external stakeholders should be managed and kept satisfied.

SUMMARY

Program governance is applied to oversee different aspects of the program, as shown in Figure 8.1, that have to be directed in order to deliver the intended benefits. It is a governance responsibility to ensure that the governance domains mentioned in this section are taken into consideration, while creating and implementing the governance framework for a particular program.

In addition to this, the reporting and disclosure from the program manager to the program governors should include reference to these assets, as appropriate, so that the oversight function of governance can proactively track any significant events within the program and can respond accordingly in order to deliver the program benefits.

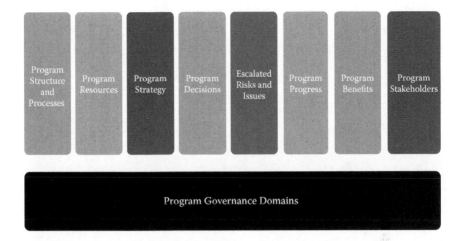

FIGURE 8.1
Program governance domains.

REFERENCES

Association for Project Management (APM). (2007). *Directing Change: A Guide to Governance of Project Management.* Buckinghamshire, UK.

Klakegg, O. J., Williams, T., Magnussen, O. M., and Glasspool, H. (2008). Governance frameworks for public project development and estimation. *Project Management Journal, 39*(1), 27–42. doi: 10.1002/pmj.20058.

Office of Government Commerce (OGC). (2007). The OGC Gateway™ process: A manager's checklist. *OGC Best Practice—Gateway to Success.* London, United Kingdom: Office of Government Commerce.

Project Management Institute (PMI). (2013a). *A Guide to the Project Management Body of Knowledge* (Fifth Edition). Newtown Square, PA: Project Management Institute.

Project Management Institute (PMI). (2013b). *The Standard for Program Management* (Third Edition). Newtown Square, PA: Project Management Institute.

Stretton, A. (2010). Note on program/project governance. [Featured Paper]. *PM World Today, XII*(I), 1–14.

Turner, J. R. and Keegan, A. (2001). Mechanisms of governance in the project-based organization: Roles of the broker and steward. *European Management Journal, 19*(3), 254–267.

9

Program Governance Functions

Program governance functions are tasks or actions which require execution in order to deliver the scope of governance discussed in the previous chapters. The functions of program governance, with the exception of a few, can be divided into two major groups based on the program life cycle. Some functions are more related to the formation and organization phase of the program, whereas others are concerned with the benefits delivery and closure phases.

This difference is evident, as during the initial phases the governance functions are more concerned with organizing the management and governance mechanism, whereas in the later phases, the focus is more toward providing support and control to the program for successful program delivery. There are, however, certain functions, such as oversight, which encompass the overall life cycle of the program and others, such as strategic alignment, that are revisited iteratively.

It is important to realize that the program governance functions are not executed independently, and there will always be a certain degree of inclusiveness among them. The distinction provided in this book is for the purpose of clarity, for pracademics to recognize the governance functions that have to be employed by the governance team during different phases of the program life cycle.

BENEFITS DEFINITION

The program governance team, executive or portfolio level entities, during the initial phase, have to define the objectives and related benefits with relative clarity and ensure the feasibility of the program. Müller (2011) refers to this concept as goal setting and mentions that the purpose of this function is "to ensure that the proposed program is feasible and its required prerequisites are either already in place or will be when needed" (p. 57). To support the organization's strategy, each program has to establish, define, and communicate its vision and goals, which should be aligned with the organization's direction.

While aligning with the work of Müller (2011), we can divide this function into three domains:

1. Identification and definition of program objectives and related benefits. This also includes a feasibility assessment of the program, in the context of the organizational culture and strategy.
2. Establishment of key performance indicators (KPIs) to measure the program's performance related to objective/benefit achievement.
3. Coordination and communication of program objectives and associated benefits, and related measures, with the stakeholders.

Defining program objectives helps the program team develop the program road map and related program management plans in a reasonably accurate manner. It also increases the stakeholders' confidence in the program's benefits as they foresee the expected program with relative transparency. Last, if not least, the measurement criteria associated with the program's objectives assist the program team to appraise the program's deliverables during its life cycle.

GOVERNANCE AND MANAGEMENT FRAMEWORK

This function is generally executed in the initial phases of the program. However, based on the feedback received or lessons learned, the governance and management framework may be revised during the course of the program, which is again a governance function.

Organizing and Structuring

From an organizing and structuring perspective, the preliminary governance entities define the program hierarchy and related roles.

The program sponsor and the program manager have to be assigned at an early stage. The attributes of the program have an impact on the selection of the program manager. Müller (2011) states that a "suitable profile of a program manager for any given program depends on the nature, size and complexity of the program" (p. 60). Thus, it is important to align the scope of the program and its context with the profile and skill set offered by the program manager. This decision is critical, as it will have a huge impact on the eventual success of the program.

Different entities that have to play a role in the governance function should be identified, and their roles and responsibilities clarified. Hanford (2004) states that "a poorly articulated management structure, overlapping roles, decision-making authorities, and roles filled by the wrong people can prevent a program from achieving sustained momentum or bog it down with endless attempts to achieve consensus on every decision." Thus, it is vital to define, and communicate relationships, between governing entities to create a synergetic governance function. These entities will include Program Management Offices (PMOs), advisory panels, the audit committee, and others, which will be discussed in the next chapter.

Executive management decides on the strategy and makes decisions; however, the governance framework assigns certain decision rights at all levels of governance (Weil and Ross 2004). The governance structure should be defined in a manner that allows decision making at the appropriate level, resulting in the reduction of negative aspects of undesirable bureaucracy, such as a decrease in program adaptability that is required for its smooth execution. Müller (2011) states that "An effective organizational structure will balance the ability to change quickly against the need for appropriate management control" (p. 59). The following questions are important when referring to decision rights:

1. What decisions must be made in order to get the maximum benefit from the investment?
2. Who should be making these decisions?
3. What will be the process for making the decisions and monitoring them?

In relation to the above, accountability at different levels has to be defined. This aspect will clarify who is responsible and accountable for the outcomes at different levels (Weil and Ross 2004). This function goes hand in hand with the decision rights, as it is better to assign accountability at the level where the decisions are being made. The accountability definition for information requirements and data, system and process ownership, tends to be an integral part of the governance function.

Process and Mechanism

From a process and mechanism perspective the governance and management framework should:

1. Provide criteria for project selection based on a program's intended benefits and organizational strategy.
2. Facilitate timely decision making.
3. Ensure consistency in decisions, actions, and results.
4. Provide a mechanism to verify program progress by identifying KPIs.
5. Provide measures to objectively judge the effectiveness of the program governance model.
6. Define the frequency and mechanism of stage gate reviews, along with the entry and exit criteria.

This board should also give attention to the risk management framework that should consider aspects related to compliance with external as well as internal standards and the establishment of risk management processes. The focus here should be on risk management mechanisms instead of addressing individual risks.

The actual management of resources is a management level responsibility; however, providing a framework is a governance function. Resource management mechanisms should optimize knowledge generation and should have practices to retain knowledge.

Finally, a quality management framework should be provided by the governance board, which will provide quality standards at the component, as well as the program level. The implementation of those standards is a program management responsibility; however, through the benefits delivery function the program's governance team reviews the program deliverables in relation to the quality standards. In case such a framework is not available at the organizational or portfolio level, the program team will have to develop the program quality plan. This plan will be approved by the program governance board.

ENSURING STRATEGIC ALIGNMENT

Programs and their component projects are considered as drivers for strategy implementation. Programs executed, within an organization, should always remain aligned to the strategic goals of the organization, or the portfolio in which they are represented. The Office of Government Commerce's (OGC) gate review 0, ongoing strategic assessment, focuses on continuous alignment of programs with the organizational strategy. APM (2007) considers lack of strategic alignment as one of the common causes of failure of such initiatives.

Organizational strategy is dynamic and changes because of various internal and external factors. This function concentrates on providing direction and guidance to the program team so that it remains aligned with the strategic vision of the organization. There should be consistent interaction between the program and the business so that any updates or changes in the business strategy, which may impact the program, can be reflected through to program objectives and related artifacts. This will ensure that the program's outcomes are consistently supporting the organization's strategic vision.

Reviews related to strategic alignment should be part of the periodic review process, where the focus should be to ensure that the program is aligned with the current business strategy, and in case any changes are required, the program governors will propose the required updates that will help in program realignment.

Thiry (2007), while proposing a program life cycle model, described another dimension to this strategic alignment. He states that the benefits delivered by the program are reviewed and are integrated in the organizational operations. This integration may result in realignment of the organizational strategy, which in turn will have an impact on the program direction.

This function encompasses the overall program life cycle. The initial strategic alignment takes place during the formation and organization phases, whereas the ongoing strategic reviews during the program life cycle ensure that the program is aligned with the current business strategy.

DIRECTING BENEFITS DELIVERY

The focus here is on the benefits generated from the program, which should be tracked and measured in terms of business outcomes and cost to benefit scales. During the program benefits delivery phase, and the program closure phases, it is important to ensure that the objectives of the constituent projects are delivered and integrated so that the defined program benefits are delivered to the beneficiary organization(s). OGC's gate review 5, readiness for service, focuses on the readiness of the organization, so that the beneficiary can gain from the program's outcomes, manage the change resulting from such an implementation, and ensure that a framework is in place to sustain the benefits (OGC 2007).

The Project Management Institute (PMI) has a full process that focuses on managing the benefits of the program (PMI 2013). Individual projects have deliverables, which are integrated at the program level to deliver the program benefits. It is a program level responsibility to define the deliverables expected from the project, ensure that they are being delivered, and integrate these deliverables so that the program can meet its defined objectives. Program governors oversee that the benefits expected from the program are realistic, and that the program is on track to deliver the expected benefits.

The program governors, through periodic reviews, have to ensure that the program's constituent projects are delivering their objectives, within the defined quality standards, constraints, and thresholds. Accordingly, it has to be ensured that these deliverables are being integrated at the program level in order to meet the program level objectives. Any deviations that are outside the threshold value are noted, decisions are made, and recommendations are provided to the program management team to bring the program back on course.

This function is more focused toward the benefits delivery and program closure phase. The emphasis during the benefits delivery phase is more toward ensuring that the program's deliverables are being created according to plan and defined standards, whereas during the program closure phase the focus is more toward the implementation and sustainment of benefits.

INTERFACING AND COORDINATION

The program governance team acts as an interface for the program with entities internal and external to the program. These entities may include units within the organization, such as other departments or functions, or external groups, such as regulatory or government authorities. The governance team provides support to the program team by coordinating and interfacing with these entities to ensure that the program remains aligned with its context.

Any conflicts between competing initiatives, such as allocation of resources to the program, are handled through this role. Unger, Gemünden, and Aubry (2012), while reviewing the role of the project portfolio management office, also discuss the coordination role. Some activities that they mention, which are aligned to the current context, include:

1. Collective collaboration with other initiatives resulting in diffusing of tensions and power struggles stemming from resource conflicts.
2. Cross-project support.
3. Cross-department coordination.

REVIEWING PROGRESS

It is important to review the overall status of the program to ensure that the program is progressing as per the approved plan, and any deviations from the plan are accounted for and approved. The program governors have to define the KPIs, which will be used to track the overall status and progress of the program from different aspects. Assessing the current state of the program, based on the measurement criteria, is one of the major responsibilities of program governance.

The program management team has the responsibility of reporting and presenting the progress on a periodic basis to the program governors. These reports should indicate the performance of the program, and its component projects, based on the KPIs defined by the program governors. They will also include reviewing the significant decisions made by the program team and its overall impact on the program. Based on the reported performance the program governors authorize and approve the program, or its components, to proceed to the next phase or to be adjusted or terminated. Even though the success of individual components is a critical aspect, it is important to understand that the program governance team should have a vision of overall program success. This vision cannot be underestimated as it ensures that the results of the components eventually deliver the value expected from the program.

Based on the program plan and progress, component projects will be initiated, closed, or terminated during the program life cycle. Such decisions also come under the influence of program governors who work with the program manager, and decide whether a particular component needs to be initiated, or another component needs to be closed while ensuring that the intended objectives are delivered. These decisions are also made while reviewing the overall program progress. Thus, the impact of these decisions is considered from an overall program benefits delivery perspective.

The program progress review encompasses the overall program life cycle, where the focus of the review will change based on the program's phase in the life cycle. This change in focus ensures that the more relevant aspects of the program are reviewed based on the current program context.

PROVIDING GOVERNANCE OVERSIGHT

Governance, as discussed earlier, is an oversight function. The program's management team requires direction and leadership throughout the program life cycle, which is provided by the program's governance team (Williams and Parr 2008). The objective of this oversight is to ensure that the program stays on track and delivers the intended benefits to the beneficiary. To meet the oversight objective, two aspects, that is, support and control, become a key governance theme. This is required because programs are of large magnitude, and all aspects of a program are not under the control of the program manager or the program management team, and they require the assistance and guidance of the program governors for successful delivery of benefits.

Programs, because of their high level of complexity and uncertainty, will not always run as planned. Governance oversight provides the overall visibility, foresight, control, and guidance that will help the program team to take preventive and corrective actions to ensure that the program delivers the desired benefits. With the main objective being delivery of program benefits, governance oversight, along with its dimensions of control and support, encompasses the overall program life cycle.

Support and Guidance

Each program needs support and guidance from the program governance team throughout its life cycle. One of the major functions of program governance is to provide the means and resources required to achieve the defined program objectives. Providing resources and ensuring resource availability is a core governance function. PMI (2013) states that the governance board has to ensure "that programs are funded to the degree necessary to support the program plan, as approved" (p. 54).

In addition, risks and issues that cannot be handled at the management level are also escalated to the governing entities for guidance and resolution. Risk management is a management level responsibility, but the board oversight should always be there to address certain organizational or escalated risks. Williams and Parr (2008) mention that one of the core functions of program governance is to provide support and facilitate issues resolution. In addition, governance entities, especially the executive sponsor, have a role to support and advocate the program in front of stakeholders.

Core aspects of support include:

1. Removing obstacles for successful program delivery.
2. Managing issues and mitigating the impact of issues that arise in the program.
3. Managing and reducing risks escalated by the team or external to the program.
4. Influencing different factors to ensure that the program constantly remains aligned with the organizational strategy.
5. Advocating and defending the program before different stakeholders, while keeping them aligned and supportive.
6. Influencing the organization to provide the necessary resources to the program team, even when such resources are not planned in advance.
7. Taking a major interest in the program by providing support and guidance to the program team in terms of decisions and direction.

To ensure consistency of decisions, actions, and results, this role also provides support to the program team by mentoring and training in terms of process implementation. Such an activity ensures that the program team is well versed in the processes and can perform their role in a more organized and consistent manner.

Monitoring and Control

The actual responsibility of monitoring and controlling the program lies with the program manager. However, the governance team has to track the progress of the program to ensure that the program is moving as per the committed plan and within the defined threshold. The reason for this function is not just to regulate the program manager's actions and program progress; it is also needed because the program manager's focus is on the program activities, whereas the governance team's vision includes organizational objectives as well as external influences, which may have an effect on the program.

It is important that the progress of the program be consistently monitored by the governance entity. Core aspects of monitoring include:

1. Conducting program governance meetings on a regular basis.
2. Conducting informal and unplanned review meetings as required.
3. Ensuring that the level and type of information in progress and status reports is available as per the program requirements.
4. Defining the frequency of program reporting and ensuring that the reporting takes place based on the communication plan.
5. Conducting planned, announced, and unannounced program audits.

Müller (2009) refers to the meetings conducted by steering groups when he describes the concept of monitoring and mentions that the frequency of such meetings is sometimes related to the life cycle phase of the initiative. This review can happen periodically or based on certain triggering events such as project authorization in a program.

From a control perspective, the level of control depends on the program's internal context, such as its status, as well as its external context, such as varying requirements of external stakeholders. Core aspects of control include:

1. Reviewing the project progress frequently and making program-related decisions.
2. Reviewing the achieved results or outcomes and making related decisions.
3. Reviewing crucial decisions made by the program team and influencing these decisions when required.
4. Reviewing resource utilization and influencing the utilization when needed.
5. Safeguarding the program from unrealistic and unstable customer expectations and requirements.

Klakegg et al. (2008), while conducting a qualitative study related to the project governance framework adopted by the public sector in the United Kingdom (UK) and Norway, also identified control aspects of governance in all large-scale initiatives that they investigated with varying degrees of intensity.

Controlling the program's progress, as well as the quality of the program's deliverables, is an important aspect of governance and is a responsibility of steering groups, such as the program governance board. The program's governance team has to exercise control when required (Williams and Parr 2008). This control is especially required when the program is not performing according to the plan and needs intervention from the program governors in terms of decision making and actions in order to bring it back on track. The program governors decide on the corrective actions and provide directions and recommendations to the program management team for implementation.

REPORTING AND COMMUNICATING

Effective reporting and communication are considered to be an important factor for improved decision making. Without a culture of honest disclosure, effective reporting cannot take place. The governance function should implement measures which assist in the disclosure of information to the relevant stakeholders. The Association for Project Management (APM) focuses on communication management and states that the criteria and content for reporting and communication, including criteria for escalation of risks and issues, should be established from the beginning (APM 2007).

Reporting of the program's progress and status to the program governors is the responsibility of the program management team; however, the program governors need to create high level or executive reports that should be disseminated, at the right time and with the right information, to executive management and relevant program stakeholders. This information delivery ensures that the executive management will be able to make accurate and timely decisions related to the program.

In some cases, there may be multiple sources of information, which may result in conflicting reports. There might be scenarios where the information collection frequency and the content of the reports are not aligned

FIGURE 9.1
Lack of trust in reports.

with the stakeholder requirements. This generally results in the lack of trust of business owners in the information obtained from the reports.

Figure 9.1 depicts the different issues that result in this dearth of confidence.

To enhance the credibility of reports created by the governance team, it is important to ensure that the reports received by them from the program team have:

1. Sources of information identified and verified.
2. Transparent reporting mechanism, including the clear process through which reports are created.
3. Conflicting information between multiple sources identified and resolved.
4. Accurate source, relevant to the information required, utilized.
5. Sufficiently specific and updated information.

In addition to the above, it is important that the communication frequency, method, and format defined by the governance team:

1. Is aligned with the stakeholder needs.
2. Is aligned with the type of information to be disseminated and with the program context.
3. Delivers information that is sufficiently specific and is updated.

Ensuring the above measures increases the confidence of stakeholders in the reports and the reporting mechanism. It is also important to understand the timing and content of information as validity of information is a key aspect. Program reporting and communication begin from the initial phases of the program; however, the frequency of reporting, as well as its content, may evolve during the program life cycle based on various factors such as program performance and other program attributes.

AUTHORIZATION AND APPROVALS

The program governance board makes decisions on every function that has been discussed in the previous section. However, there are certain specific decisions that come under the direct responsibility of the program board. These decisions define the direction of the program and can significantly affect the program's progress and prospects.

While aligning with PMI (2013), the following are some of the decisions that are undertaken by the program governance board:

1. The program governance board authorizes the formation and organization of a program. This is done through review and approval of the program's business case and the program charter. Approval of the program charter provides a formal go-ahead to move to the organization phase of the program.

2. The program road map, approach, and program management plan are developed by the program management team, with support from other stakeholders. However, in order to execute those plans, a formal authorization has to be provided by the program governance board. In addition, any significant changes to the plan beyond the threshold have to be approved by the governance board through a formal change management process.

3. Any component that is initiated in the program has to be approved by the governance board. Component initiation acts as a checkpoint where the program's alignment with organizational strategy is verified. It is important for the governance board to review the component initiation request, which is generally submitted by the program manager, so that it can revise the program governance framework to cater to the requirements of the new component and can secure resources for the same.

4. Any component closure, either through completion or termination, has to be approved by the program governance board. The program sponsor may also approve component closure.

5. Once all expected benefits are delivered, or the need for the program ceases to exist, the program is closed. The formal closure of the program is approved by the governance board. Postprogram appraisals are conducted by the governance board to review the overall benefits

delivered and to assess the level of program success based on the KPIs defined during the prior stages. In case of termination, the reasons behind termination are reviewed to ensure their validity. Based on the closure approval, formal activities related to program closure are executed.

Other decisions that come under the domain of the governance team include decisions related to escalated issues and risks, approval on significant deviations from plans, and additional resource/funding requirements. The governance board should have a well-defined decision-making mechanism in place in order to support the program's complex, adaptable, and uncertain nature.

――――――

SUMMARY

The governance functions defined in this chapter address different governance domains that were discussed in the previous chapter. Table 9.1 provides a mapping between governance functions and the governance domain that they typically address. This linkage can act as a guideline for pracademics to comprehend and design an efficient and effective governance framework that addresses all areas that come under the obligation of the program governance team. It should, however, not be considered as a holistic or all-inclusive table, and can be altered based on certain organizational needs and context.

It is important to reiterate that certain governance functions are more relevant to selected stages of the program life cycle. There are, however, other governance functions, such as governance oversight, that are required throughout the program life cycle. The diagram in Figure 9.2 provides a conceptual mapping of the program life cycle and the program governance functions. This generic mapping can be applied to most programs, most of the time.

The responsibility of governance eventually lies with the board of directors; however, the board does not perform all these activities by itself. Instead it generally creates certain roles and institutions and assigns responsibilities to them to perform the governance function. This results in the creation of roles, such as executive sponsors and institutions, including but not limited to the PMO, steering committees, portfolio(s),

TABLE 9.1

Linking Program Governance Functions and Domains

Governance Function	Governance Domain(s) Addressed
Benefits definition	Program benefits, program strategy
Governance and management framework	Program structure and processes
Ensuring strategic alignment	Program strategy
Directing benefits delivery	Program resources, program strategy, program decisions, program benefits
External interfacing and coordination	Program resources, program benefits, program stakeholders
Reviewing progress	Program resources, program decisions, escalated risks and issues, program progress, program benefits
Providing governance oversight	Program structure and processes, program resources, program decisions, escalated risks and issues, program progress, program benefits
Reporting and communication	Program decisions, escalated risks and issues, program progress, program benefits, program stakeholders
Authorization and approvals	Program structure and processes, program resources, program strategy, program decisions, escalated risks and issues, program benefits

and regimes. They align and work together to fulfill the organizational requirements from program governance. The next chapter will provide a detailed review of different governance roles.

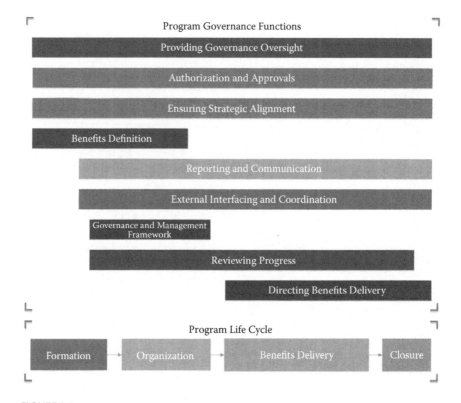

FIGURE 9.2

Mapping governance functions with a program life cycle.

REFERENCES

Association for Project Management (APM). (2007). *Directing Change: A Guide to Governance of Project Management*. Buckinghamshire, UK.

Hanford, M. F. 2004. *Program Management: Different from Project Management*. IBM. Retrieved from http://www.ibm.com/developerworks/rational/library/4751.html on March 13, 2013.

Klakegg, O. J., Williams, T., Magnussen, O. M., and Glasspool, H. (2008). Governance frameworks for public project development and estimation. *Project Management Journal, 39*(1), 27–42. doi: 10.1002/pmj.20058.

Müller, R. (2009). *Project Governance*. Surrey, UK: Gower.

Müller, R. (2011). Project governance. In Pinto, J., Morris, P., and Söderlund, J. (Ed.), *Oxford Handbook of Project Management*. Oxford, UK: Oxford University Press.

Office of Government Commerce (OGC). (2007). The OGC Gateway™ process: A manager's checklist. *OGC Best Practice—Gateway to Success*. London, United Kingdom: Office of Government Commerce.

Project Management Institute (PMI). (2013). *The Standard for Program Management* (Third Edition). Newtown Square, PA: Project Management Institute.

Thiry, M. (2007). Program Management: A strategic decision management process. In Morris, P. and Pinto, J. K. (Ed.) *The Wiley Guide to Project, Program, and Portfolio Management*. Hoboken, NJ: John Wiley & Sons.

Unger, B. N., Gemünden, H. G., and Aubry, M. (2012). The three roles of a project portfolio management office: Their impact on portfolio management execution and success. *International Journal of Project Management*, *30*(5), 608–620.

Weil, P. and Ross, J. W. (2004). *IT Governance: How Top Performers Manage IT Decision Rights for Superior Results*. Boston, MA: Harvard Business School Press.

Williams, D. and Parr, T. (2008). *Enterprise Programme Management—Delivering Value* (p. 303). New York: Palgrave Macmillan.

10

Program Governance Institutions and Roles

One of the key factors in having an efficient program governance framework is the optimum organization and structuring of the program's governance team. The roles and entities for governance and the governance reporting structure may differ from organization to organization, and the type of initiatives that have to be governed. The structure needs to be designed carefully while keeping all the factors in perspective in order to fulfill the governance needs of the program. The governance structure should support the program and facilitate delivery. It should not add additional layers of mindless bureaucracy that result in delayed decisions and subsequent actions.

An example of the design and adaptability of a governance structure, based on project requirements, is the North-South Metro Line project (Staal-Ong and Westerveld 2010). The project delivery structure evolved from a hierarchical model to a more independent leaner model. During the first phase of the project, the governance structure had relatively unclear responsibilities and authority between Projectbureau and Adviesbureau, the two entities responsible for project delivery. In addition, the dependency on different government entities increased the bureaucracy. This resulted in increased decision-making and reaction times. The structure also created principal–agent relationship problems between Projectbureau and Adviesbureau. The agent (Adviesbureau) was more concerned with delivering the project within the financial constraints instead of focusing on the overall project objectives.

However, the project delivery organization adapted and evolved based on recommendations or interventions of external committees. The responsibilities between Projectbureau and Adviesbureau were gradually integrated resulting in the synchronization of objectives and elimination of agency problems. It was also planned that from 2010 the project delivery organization would become an independent entity with full responsibility and authority to deliver the project objectives. This resulted in a leaner and efficient project delivery and governance structure which reduced the bureaucratic environment.

This chapter provides an overview of different entities and roles that are involved in the program governance. The title of governing entities can differ from organization to organization, however, the roles identified here can fulfill the governance needs of a typical program.

THE BOARD OF DIRECTORS

The board of directors (BoD) is responsible for defining and driving the organizational strategy, along with the management team. As part of the strategy formulation, the board decides on the allocation of funds to different types of initiatives that have to be sanctioned in order to achieve the organizational objectives. If the organizational strategy is delivered through a portfolio, this decision acts as a guiding principle at the portfolio level where the portfolio management team decides on initiating/continuing/suspending projects and programs (Müller 2011). The BoD functions as a governance entity responsible for the oversight of several projects and programs.

From a governance perspective, it is the board's obligation to set up the overall governance mechanism in the organization. BoD delegates program governance duties by forming entities or roles, which oversee a specific program and report to the BoD. These entities have some sort of independence and have specific responsibilities to fulfill within the overall governance function. The number and responsibilities of these committees may differ based on the governance needs and focus. The decision to set up steering committees, program governance boards, and other governance entities is within the purview of the board of directors.

The nominated BoD members, along with the program governance board, define and communicate the roles of all entities involved in the program governance function. These members may also ensure that the governance framework of the program is aligned with the overall governance principles and requirements of the organization. The creation of an appropriate governance environment in which programs and projects can flourish and succeed comes under the direct responsibility of the BoD.

TABLE 10.1

Linking the Board of Directors and Its Core Governance Functions

Governance Institutions and Roles	Core Governance Function(s)
Board of directors	Benefits definition
	Governance and management framework
	Ensuring strategic alignment
	Reviewing progress
	Authorization and approvals

Table 10.1 shows the typical functions of governance where the BoD contributes significantly or which are under its direct responsibility.

PORTFOLIO MANAGEMENT

To ensure that projects and programs are strategically aligned and have some relationship or dependencies among them, it is best to align them under strategic portfolios. All portfolios have to be managed to achieve the strategic objectives with which the portfolio is aligned. This approach includes defining, selecting, aligning, as well as realigning, coordinating, and terminating portfolio components to achieve the higher-level strategic objective assigned to it.

To refresh our memories, I will revisit how researchers and professional associations have defined portfolio and portfolio management.

Turner and Müller (2003) define the portfolio as an "organization (temporary or permanent) in which a group of projects are managed together to coordinate interfaces and prioritize resources between them and thereby reduce uncertainty" (p. 7).

PMI (2013b) defines the portfolio as "a component collection of programs, projects, or operations managed as a group to achieve strategic objectives. The portfolio components may not necessarily be interdependent or have related objectives" (p. 3). Portfolio management in turn is defined by PMI (2013b) as "the coordinated management of one or more portfolios to achieve organizational strategies and objectives" (p. 5).

OGC (2009) refers to the portfolio as the alignment and compilation of all "programmes and stand-alone projects being undertaken by an organization, a group of organizations, or an organizational unit" (p. 308).

APM (2009) defines a portfolio as a "grouping of projects, programmes, and other activities carried out under the sponsorship of the organization" (p. XV). APM (2009) defines portfolio management as "the selection and management of all of an organisation's projects, programs, and related business-as-usual activities taking into account resource constraints. A portfolio is a group of projects and programs carried out under the sponsorship of an organisation. Portfolios can be managed at an organisational, program or functional level" (p. 8).

All of the above definitions mention that the portfolio is a collection of projects, programs, and related work; however, PMI's definition provides a reference to alignment between the portfolio and strategic business objectives; whereas APM and OGC do not relate to any such linkage.

The program and projects within a portfolio are temporary; however, portfolios are generally permanent organizations within parent organizations, which are focused toward certain organizational strategic goals or objectives. The main objective of portfolio management in the governance of programs is related to governing the alignment, prioritization, selection, and termination of initiatives based on the organizational strategy.

Phase gate review, which is a governance mechanism for programs initiated within a portfolio, should be carried out in the context of the related portfolio. This approach makes portfolios an overarching entity that governs programs. The overarching portfolio receives communication and reports from the program. The portfolio team provides direction to the program in terms of organizational strategy, support in resource allocation, and component prioritization.

The typical functions of governance where the portfolio management team contributes significantly or which comes under its direct responsibility are shown in Table 10.2.

EXECUTIVE SPONSORS

Executive sponsors act as a link between the permanent organization and programs, where they provide governance oversight required by the

TABLE 10.2

Linking Portfolio Management and Its Core Governance Functions

Governance Institutions and Roles	Core Governance Function(s)
Portfolio management	Benefits definition
	Ensuring strategic alignment
	Reviewing progress
	Providing governance oversight
	Authorization and approvals

parent organization and obtain support for the program from the corporation. The executive sponsor is generally a senior executive who leads a business unit, which is a major program beneficiary, and has a role in directing the organization and its investment decisions. There is also a possibility that multiple executives will share this role, however, in such cases there would still be a senior executive sponsor who will act as a final decision maker (Hanford 2005). He or she also provides the resources to the program and its constituent projects to ensure the successful delivery of program benefits.

Crawford et al. (2008) suggest that sponsorship is one of the key aspects of governance, which acts as the main linkage between corporate governance and the governance of projects/programs. Responsibilities for this role are derived by the corporate governance needs to provide the governance oversight on the project/program and, at the same time, support the project/program from contextual issues that might impact the project/program deliverables.

Considering sponsorship to be a key factor for program success, sponsorship should come from the top management level. A program needs sponsorship from people/entities who have a position of authority in the organization. This is because organizations may have stakeholders whose interests might be negatively affected by the program, and who may try to influence the program outcomes. However, Crawford et al. (2008) further suggest that the sponsor should be available for the program as required and should be someone who is accessible when needed.

The gathering of political support and fighting for the programs are considered to be important aspects of the sponsorship role. Sponsors need to work with various stakeholders and manage their expectations. At the same time they need to guide the program's management team so that they can deliver the program benefits that are aligned with the stakeholder's expectations.

Crawford et al. (2008) developed an understanding of the sponsorship role and its impact on the project/program success through a study which involved nine organizations, 36 projects/programs, and five geographical regions: Australia, China, Europe, North America, and South Africa. Two important sponsorship roles that have been discussed by Crawford and her fellow researchers include:

1. Support, which is influenced by the project/program needs.
2. Governance, which is influenced by the corporate governance needs of the parent organization.

The focus of the sponsor should shift from support to governance and vice versa as the context of the project/program changes, that is, if a project/program is performing poorly, or if there are some external regulatory requirements that need to be fulfilled, the focus moves from support to governance. If the parent organization is not supporting the project/program or if there are negative stakeholders causing problems in the execution, then the sponsor should focus more on the support aspect (Crawford et al. 2008).

Some of the key characteristics of sponsors include (Crawford et al. 2008):

1. Effective communication skills
2. Commitment to the project and its objectives
3. Seniority of position in the organization
4. Availability of the sponsor for the management team

The executive sponsor "owns" the program and works closely with the program manager to ensure the program is progressing according to plan and is receiving the appropriate support from the organization. OGC proposed the role of senior responsible owner (SRO), who is considered as the executive sponsor for projects and programs (OGC 2007). OGC, while referring to the gate review process, mentions that the ownership of the implementation of recommendations for improvement lies with the SRO. The SRO must be a senior person who will ensure that the program meets the required objectives and that the execution is managed properly.

Other countries have implemented similar roles in the way their government is set up. Senior manager ownership and leadership can be considered as one of the criteria for the successful delivery of programs and

TABLE 10.3

Linking the Executive Sponsor and Its Core Governance Functions

Governance Institutions and Roles	Core Governance Function(s)
Executive sponsor	Benefits definition
	Governance and management framework
	Ensuring strategic alignment
	Directing benefits delivery
	External interfacing and coordination
	Reviewing progress
	Providing governance oversight
	Reporting and communication
	Authorization and approvals

projects. APM defines the lack of clear senior management involvement as a common cause of failure of projects and programs (APM 2007).

Some researchers have differentiated between the roles of the owner and sponsor. Turner (2006a) mentions that the owner "provides the resources to buy the asset and will receive benefits from its operations" (p. 3), whereas the sponsor "will channel the resources to the project on the owner's behalf" (p. 3).

Sponsorship is identified as a leadership role within project/program context. OGC (2007), while referring to its gate review process, mentions senior manager ownership and leadership as one of the criteria for successful delivery of projects and programs. PMI (2013a) enforces the importance of the sponsorship role, and mentions that the "program sponsor is the individual responsible for championing the application of organizational resources to the program and for ensuring program success" (p. 65).

The role of executive sponsor is central to the governance of programs, thus the executive sponsor contributes significantly by assuming responsibilities of all governance functions (Table 10.3).

THE PROGRAM GOVERNANCE BOARD

The concept of the program governance board/steering committee comes from the corporate governance function where the board of directors sets up such groups to work closely with the management team to support the execution and management of initiatives (ITGI 2003). Unlike top-level strategy committees, these steering committees or program boards have a more hands-on approach as they are involved more closely with the program management team in order to guide and support them in their work.

Programs, because of their magnitude, typically impact multiple business units and stakeholders. The program board provides these stakeholders with a platform to sit together to provide direction to the program during its life cycle. Hanford (2005) mentions that the role of these boards is mostly to direct and provide guidance to the program. Decisions related to the governance mechanism, such as the frequency and content of audit activities, reporting frequency and format, stage gate reviews, and their objectives are made by the board.

Issues, risks, and decisions escalated from the program level reach the program board where different representatives work together to reach an agreement, which should be approved with consensus. Additionally, the board is also involved in program performance reviews and audits. It also approves initiation, closure, and transition of program components as well as the overall program. The program governance board, along with the executive sponsor, provides the required resources to the program, reviews the program progress, and approves changes. Such boards are headed by a program board director, who typically is the executive sponsor, a role that was discussed under "Executive Sponsor," earlier in this chapter.

PMI (2013a) recommends that the board should be staffed with members who:

1. Have strategic insight and decision-making authority in terms of setting up the program goals, strategy, and operational plans.
2. Have an influence on provisioning of resources for the program.
3. Have strategic insight, technical knowledge, functional responsibilities, operational accountabilities, responsibilities for managing the organization's portfolio, and/or abilities to represent important stakeholder groups.

PMI (2013a) suggests that members who have roles aligned with the above domains can improve the effectiveness of the board resulting in improved program performance. This is because the program governance board will be equipped with the authority and tools to efficiently address issues or questions that may arise during the program life cycle.

TABLE 10.4

Linking the Program Governance Board and Its Core Governance Functions

Governance Institutions and Roles	Core Governance Function(s)
Program governance board	Benefits definition
	Governance and management framework
	Ensuring strategic alignment
	Directing benefits delivery
	External interfacing and coordination
	Reviewing progress
	Providing governance oversight
	Reporting and communication
	Authorization and approvals

Based on the type and complexity of the program being governed, the program might report to multiple governance boards. This is especially true in cases where the program is being jointly sponsored and overseen by multiple entities such as a consortium between government and private entities or joint implementation by multiple vendors. In such cases, the roles, responsibilities, and authority of each overseeing board must be established at the time of inception.

The program governance board contributes significantly by assuming responsibilities of all governance functions (Table 10.4).

THE PROGRAM MANAGEMENT OFFICE

The program management office (PMO) plays different roles in different organizations. The particular role of the PMO depends on the organizational requirements of such an entity. PMOs can play a support function where they set up the program management standards and train and guide the program management teams to implement those standards. This is a tactical role where the focus of the PMO is governance through behavioral control.

PMOs typically include functions that are related to a single program/project; however, the PMO can become more effective if it includes functions that impact program/project management at the organizational level. Jainendrukumar (2008) mentions that the typical functions of a PMO can include:

1. Practice management, which is related to setting up the processes and practices for program/project management and governing its implementation.
2. Infrastructure management, which includes functions related to program/project governance and assessment. This also includes providing tools, facilities, and equipment to the team.
3. Resource integration, which is focused toward providing support to the team in terms of resource pool management, team development, and training.
4. Technical support management, which includes functions related to mentoring of teams, assistance in planning, audits, and providing support in case of recovery.
5. Business alignment, which is a more strategic function and is related to management of the portfolio of programs and projects as well as managing external relations on behalf of the programs and projects.

There may be other cases where PMOs have a more operational role to play and act as a reporting entity where they are responsible for:

1. Receiving reports from individual projects within the program.
2. Analyzing and compiling reports at the program level.
3. Reporting the program progress to various stakeholders.

In some organizations PMOs take the role of a center of excellence (CoE), which has the responsibility to ensure "a high level of consistency and professionalism in the management of their programs" (PMI 2013a, p. 64.) The focus here is on implementing governance through behavioral control as the PMO provides:

1. Staff highly trained in the program management domain.
2. Practices of applying program management processes and standards.
3. Centralized, consistent program management expertise to a portfolio of initiatives.

Müller (2011) also refers to PMOs playing the strategic and business alignment role as strategic PMOs, which are engaged in the "stewardship of portfolios of projects" (p. 308). Such PMOs work with the portfolio manager in terms of providing reports for more effective decision making

TABLE 10.5

Linking the PMO and Its Core Governance Functions

Governance Institutions and Roles	Core Governance Function(s)
Program management office (PMO)	Governance and management framework
	Directing benefits delivery
	Providing governance oversight
	Reporting and communication
	Authorization and approvals

on projects and programs; thus, their focus is governance through outcome control (Müller 2011).

Table 10.5 shows the typical functions of governance where the PMO contributes significantly or which comes under its direct responsibility.

ADVISORY PANELS

An interesting concept to govern programs is the creation of advisory panels. Pells (2007) refers to these panels as project management advisory panels, and mentions that such panels should be comprised of individuals who are experts in the field program/project management and have experience relevant to the domain of the program under consideration. This group of experts can assist the program management team in resolving risks and issues, and can provide guidance and direction to the program in terms of costing, scheduling, and general management of the program. There should be a consistent interaction between the program management team and the advisory panel through periodic program reviews or on a consultative basis.

Based on my consulting experience with medium- to large-size organizations and programs, advisory panel members should have:

1. Expertise and significant experience in project, program, and portfolio management and related practices.
2. Proficiency in the domain related to the scope of the program, thereby providing technical advice where necessary.
3. Experience in delivering similar initiatives in the past, thereby ensuring that the lessons they have learned from their past experiences can be utilized in the current initiative.

TABLE 10.6

Linking Advisory Panels and Their Core Governance Functions

Governance Institutions and Roles	Core Governance Function(s)
Advisory panels	Benefits definition
	Governance and management framework
	Ensuring strategic alignment
	Directing benefits delivery
	Reviewing progress
	Providing governance oversight
	Authorization and approvals

4. A consulting and advisory mind-set, thereby ensuring that they do not get involved in day-to-day activities of the program.

In addition to the above, it would be ideal to involve practitioners and academics who are considered as thought leaders in the relevant field so that the program governance and management team respects their opinions and consults with them on a regular basis. This involvement helps the program to stay on course in order to avoid the need for later corrective action.

Table 10.6 shows the typical functions of governance where advisory panels contribute significantly or which come under their direct responsibility.

AUDIT COMMITTEES

Programs are complex entities that have a high level of visibility. Organizations, through the governance mechanism, define standards and processes that ensure programs deliver consistent results. To warrant that a program follows the defined practices, including management processes, organizations conduct audits at different stages of the program life cycle. Other areas of program audits include:

1. Program financials
2. Program quality
3. Program documentation

TABLE 10.7

Linking the Audit Committee and Its Governance Functions

Governance Institutions and Roles	Core Governance Function(s)
Audit committee	Governance and management framework
	Directing benefits delivery
	External interfacing and coordination
	Reviewing progress
	Reporting and communication
	Authorization and approvals

To ensure that programs are ready for audits, the program governance board may form an audit committee. This committee is responsible for preparing the program for performance and quality assurance reviews. Additionally, it also ensures that the program consistently complies with the organizational rules and external codes of conduct. PMI (2013a) mentions that the audit committee assumes the "responsibility for creating or employing organizational infrastructure to support the effective audit of programs, such as an information repository" (p. 65).

Such committees reduce the burden of audits, and related activities, by providing the program with the required resources, training the program team to comply with the regulations, mentoring, and supporting them in terms of process implementation, supporting them in financial management, and providing them with the tools to help them manage program documentation. This assistance ensures that the program remains prepared for audits, whether they are preplanned, announced, or sudden.

Table 10.7 shows the typical functions of governance where the audit committee contributes significantly or which comes under its direct responsibility.

GENERIC ROLES—THE BROKER AND STEWARD

Turner and Keegan (2001) applied the lens of transaction cost economics (TCE) while identifying the governance mechanism, structure, and roles in project-based firms. They mention that the buyer and seller relationship is present in all transactions within project-based firms where the buyer or seller can be internal or external to the organization. They define a project-based organization as one in which the "majority of products are made or services are against bespoke design for customers" (Turner and Keegen, 2001, p. 256). Such organizations can have any or all of the following types of undertakings:

1. They undertake multiple projects for multiple clients, which is basically a portfolio of projects.
2. They undertake a program of projects for customers in which all projects are in the end contributing to the program objectives.
3. They undertake large projects for clients such as major construction initiatives.
4. They are a start-up company created to develop a product or service. They ensure that the company and the product it is creating are a project on its own.

To govern such an undertaking, the roles of the broker and steward are present, and these are at the core of governance structure of project-based firms (Turner and Keegan 2001). As there are two parties involved in projects, both roles focus on different aspects of governance.

The broker is a client-facing role who maintains the external relationships with the customer, whereas the steward aligns and assigns resources internally and makes sure that the resources are available when the initiative requires them. The broker also has a key role to play in getting new customers and so he or she is also involved in governance activities before the initiation and after the closure. This can include the identification of new customers, getting involved in bidding processes, maintaining customer relations, and ensuring that the customer is satisfied so that new work can be acquired from the same customer (Turner and Keegan 2001). Both roles interact with each other as well to create the link between both the internal and external perspectives.

The roles of the broker and steward are present in all project/program scenarios with varying degrees of responsibility and with different titles assigned to the roles. From a TCE perspective, this difference in responsibilities is present in order to structure a governance regime aligned with the initiative (transaction) being undertaken so that economization can be achieved (Turner and Keegan 2001). In case of direct contract needs between the two parties, it might not be economical to create additional levels of bureaucracy, and the owner may take care of these responsibilities for both the broker and steward. However, in the case of strong communication needs, where there are multiple relationships involved, these responsibilities are assigned to the broker and steward.

Turner (2006b) revisited the concept of the broker and steward while working toward creating an integrated theory of project management and mentions that the broker "is a person who defines the objectives of the project, the desired outcome (benefit) and defined output (deliverable, facility or asset)" (p. 93); whereas, the steward "is the person who defines the means of achieving the objectives" (p. 93). The broker remains committed to seek the best interest for the owner and works with the owner to define the required outputs and outcomes; whereas the steward has to be realistic in terms of objectives and scope, and works with the broker to identify means, resources, and work required to deliver the output. Both these roles work together to achieve a solution accepted by all to achieve the defined outcomes.

One can relate the role of the broker to account managers or engagement managers who have the responsibility of handling customer expectations and requirements while collaborating with the project delivery team to ensure successful execution of the project. The steward, an internal looking role, can be aligned to the role of project director, who has the responsibility of providing the resources to the project team and governing the initiative internally.

SUMMARY

In this chapter, we have discussed the different roles and governance entities that organizations create to govern large-scale initiatives. The actual configuration of the governance structure will differ based on multiple factors such as the type of program being governed and the organizational context, as well as specific needs of the stakeholders involved. In addition, the responsibilities of the entities defined above are not fixed, may overlap, and may differ from organization to organization.

In some organizations, programs act as a governance entity for projects, whereas portfolio management acts as a governance entity for projects and programs. The program governance board is chaired by the executive sponsor, who is ultimately responsible for the success of the initiatives. The program governance board shares the governance responsibility with the portfolio manager, especially in cases where issues and decisions cannot

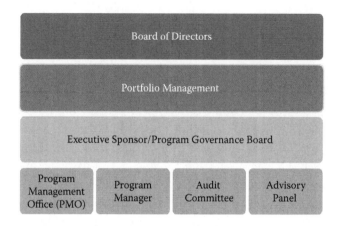

FIGURE 10.1
Program governance structure.

be resolved at the governance board level. Audit committees, advisory panels, and the PMO control different aspects of the program and provide support as required by the program governance board.

Based on my experience with certain government and nongovernment entities, a typical program governance structure, based on the entities defined above, is shown in Figure 10.1.

REFERENCES

Association for Project Management (APM). (2007). *Directing Change: A Guide to Governance of Project Management.* Buckinghamshire, UK.

Association for Project Management (APM). (2009). *APM Body of Knowledge* (Fifth Edition). Buckinghamshire, UK.

Crawford, L., Cooke-Davies, T., Hobbs, B., Labuschagne, L., Remington, K., and Chen, P. (2008). Governance and support in the sponsoring of projects and programs. *Project Management Journal, 39*(1), 43–55. doi: 10.1002/pmj.20059.

Hanford, M. F. 2005. *Defining Program Governance and Structure.* IBM. Retrieved on March 13, 2013 from http://www.ibm.com/developerworks/rational/library/apr05/hanford/.

IT Governance Institute (ITGI). (2003). *Board Briefing on IT Governance.* Illinois: IT Governance Institute.

Jainendrukumar, T. D. (2008). The project/program management office (PMO). [Feature Paper]. *PM World Today, X*(I), 1–9.

Müller, R. (2011). Project governance. In Pinto, J., Morris, P., and Söderlund, J. (Ed.), *Oxford Handbook of Project Management.* Oxford, UK: Oxford University Press.

Office of Government Cormmerce (OGC). (2007b). *Managing Successful Programmes (MSPTM).* Norwich, UK: The Stationery Office (TSO).

Office of Government Commerce (OGC). (2009). *Managing Successful Projects with PRINCE2*. London, United Kingdom: The Stationery Office (TSO).

Pells, D. L. (2007). Expert Advisory Panels for Project Management Governance and Oversight. *Paper presented at the 1st UTD Project Management Symposium*, Plano, Texas.

Project Management Institute (PMI). (2013a). *The Standard for Program Management* (Third Edition). Newtown Square, PA: Project Management Institute.

Project Management Institute (PMI). (2013b). *The Standard for Portfolio Management* (Third Edition). Newtown Square, PA: Project Management Institute.

Staal-Ong, P. L. and Westerveld, E. (2010). The North-South Metro Line, managing in crowded historic Amsterdam. In Turner, J. R., Huemann, M., Anbari, F. T., and Bredillet C. N. (Ed.), *Perspectives on Projects*. New York: Routledge.

Turner, J. R. (2006a). Towards a theory of project management: The nature of the project. *International Journal of Project Management, 24*(1), 1–3.

Turner, J. R. (2006b). Towards a theory of project management: The nature of project governance and project management. *International Journal of Project Management, 24*(2), 93–95.

Turner, J. R. and Keegan, A. (2001). Mechanisms of governance in the project-based organization: Roles of the broker and steward. *European Management Journal, 19*(3), 254–267.

Turner, J. R. and Müller, R. (2003). On the nature of project as a temporary organization. *International Journal of Project Management, 21*(1), 1–8.

11

Program Governance Mechanisms

In my last assignment, I was part of a program review committee where my role was to evaluate the management and governance approach being proposed by the vendor. It was a turnkey initiative to combine multiple entities of the government department in order to remove the duplication of efforts, unify shared services under excellence centers, and improve organizational efficiency. The program had a five-year timeline. However, while reviewing the governance approach, the program had four stage gate reviews. A five-year program and only four stage gate reviews?

Upon inquiry, the vendor responded that it had used a similar approach in an earlier project. The earlier project had a nine-month timeline and a singular focus on technology implementation. The approach did not make sense to me nor to my committee members, thus we recommended major changes in this governance mechanism. The new approach included stage gate reviews at major deliverables, monthly health checks, and strategy reviews. Finally, major reviews at the end of every quarter were incorporated. There was an agreement that the governance approach would be revisited at the end of the quarter and would be adjusted based on the program context.

Based on my experience and guidance and learning from mentors, I divide program governance mechanisms broadly based on two themes:

1. Planned or on-demand events
2. General review or specialized reviews

The following sections define governance mechanisms that are employed by organizations around the world, with different configurations, to effectively execute governance functions discussed in previous chapters.

STRATEGIC REVIEWS

One of the core aspects of governance is to keep the program aligned with the organizational strategy from formulation to closure. Thus, strategic alignment becomes a key area which requires specific focus and may require a separate forum for review. Strategic review sessions ensure that the program maintains a connection to the organizational strategy throughout its life cycle.

Hanford (2005) mentions that organizational strategy is not something static as it keeps on evolving based on multiple internal and external factors. Strategic review sessions provide the platform which maintains a constant communication channel with the organizational strategy. This link ensures that the program makes the adjustments required to keep itself aligned with the larger organizational context.

This mechanism should be put in place from the program's formulation phase and should be executed throughout the program's life cycle. The focus of these sessions differs based on the program life cycle stage. Gate review 0, "Ongoing Strategic Assessment," which is one of the six gate reviews proposed by the Office of Government Commerce (OGC), focuses on the strategic alignment of the programs and component projects with the organizational strategy (Hanford 2005; OGC 2007).

During the formulation and organization stage, the sessions focus on ensuring that the program performs an initial alignment with the organizational strategy and clarifies its goals and objective. Multiple sessions may have to be performed as the clarity can only be achieved through iterative discussions. The input from the program team can also act as a factor that influences the organizational strategy. The main objective of this session is to ensure that the program plans consider the organization's business strategy as a core input. Hanford (2005) states that the session carried out at this stage can have the following outcomes:

1. Program goals and strategy
2. Program capital and expenses budget
3. Program benefits definition
4. Program outline
5. Candidate projects identification
6. Program mobilization plan
7. Consulting and staffing agreements

TABLE 11.1

Strategic Review Session—Attributes

Session type	Preplanned sessions based on dates, milestones, or events
Session focus	Specific focus on program's strategic aspects.
Target participants	BoD members, program's management team, executive sponsor, program governance board members, portfolio management team, and advisory panel

During the benefits delivery stage period reviews are performed. These reviews focus on consistent alignment of the program's plan with the organizational strategy. These reviews can happen in conjunction with the stage gate reviews or other program events such as a request for component initiation, component transition, or need for integration of component deliverables. Hanford (2005) mentions that during these sessions "reviewers can compare the program's current state and results against the then-current business strategy, and propose needed adjustments."

Table 11.1 summarizes the different attributes of strategic review sessions.

STAGE GATE REVIEWS

One of the mechanisms to have in a consistent governance regime is to implement a gate review process at key decision points or phase-end in the program. PMI mentions the stage gate review as a key mechanism to provide governance oversight (PMI 2013). The objective of stage gate review is to measure the program's progress from different perspectives and to make key decisions related to the future direction of the program. These reviews generally occur at the end of each phase of the program or at significant events. The schedule, objectives, and content of stage gate reviews are defined in the program governance plan.

The actual content of every review may vary, however, in alignment with the recommendations of PMI (2013). The objective of such reviews is to:

1. Bring all stakeholders on board, and get them updated on multiple aspects of the program.
2. Ensure the program components are aligned with the goals of the program, whereas the program is aligned with the intended strategic objectives.

3. Review and measure the program's progress and provide feedback regarding the same.
4. Review the benefits achieved at this stage and their transition to operations. This ensures conducting a quality assessment of the benefits delivered.
5. Ensure that the program activities are being performed utilizing the defined procedures and practices.
6. Review resource utilization and provide commitment for the future program needs.
7. Review risks and issues associated with the program, and ensure that the overall risk exposure of the program is within acceptable range or provide support in terms of risk response.
8. Validate the program's position from an organizational strategy perspective.
9. Review significant decisions made at the management level while understanding the decision-making process and implications of those decisions.
10. Ensure stakeholder satisfaction with current program performance.
11. Make significant decisions related to the program and its component projects that might impact the program's future direction.
12. Decide on continuation of the program to the next phase, suspension or termination of the program based on the current progress, program context, and the organization's current strategy.
13. Decide on the future course of the program such as approving components for subsequent initiation.

Gate reviews generally occur at predefined milestones, such as the completion of a program phase, completion of a component project, a request to initiate a new component, or other major events. The objective of such reviews is to monitor the program's progress and the current status of benefits delivery and outcomes so that significant decisions can be made, which will influence the program future and direction. These reviews provide an opportunity for the program governance board to either support the program's current direction or to provide directives that will enhance the program's ability to deliver the intended benefits (PMI 2013).

OGC's gate review process is a governance mechanism, which is comprised of six gate reviews (gate review 0 to gate review 5) conducted at key decision points by experienced professionals who are not part of the project/program team (OGC 2007). These six gate reviews and their objectives are defined below (OGC 2007):

1. Gate review 0, which is "Ongoing Strategic Assessment," is specific to programs, and is conducted throughout the life cycle of the program at key decision points. This review focuses on the strategic alignment of the programs, component projects, and their progress. The objective of this review is to ensure that the need for the program still exists and that the program will meet its desired outcomes.

2. Gate review 1, which is "Business Justification," is a project-level review, and is conducted prior to project initiation after the business case is ready for review. The objective is to validate the business justification of the project before approving preparation of the development proposal.

3. Gate review 2, which is "Delivery Strategy," is a project-level review, and is conducted after the delivery strategy is developed. The objective is to verify the delivery approach and to make sure that viable implementation plans are in place. Vendors are only approached for delivery execution after the project has passed this review.

4. Gate review 3, which is "Investment Decision," is a project-level review, and is conducted after proposals are received from the vendors. The objective of this review is to revalidate the full business case and make sure that governance arrangements are in place before the investment decision is made and resources are committed.

5. Gate review 4, which is "Readiness for Service," is a project-level review, and is conducted after the project is ready to be commissioned for service. The objective here is to verify the readiness of the organization to implement, manage change, and govern the service after delivery.

6. Gate review 5, which is "Operations Review and Benefits Realization," is a program-level review, and is conducted at regular intervals throughout the life cycle of the service starting from its commissioning until the service is decommissioned. The objective of this ongoing review is to ensure smooth operations and maximum benefit realization from the service.

Such a governance mechanism ensures that the right programs and their projects are chosen, the projects remain aligned with their objectives

TABLE 11.2

Stage Gate Review—Attributes

Session type	Preplanned sessions based on dates, milestones, or events
Session focus	General view of different aspects of the program such as progress, risks, issues, decisions, and other necessary areas
Target participants	BoD, program management team, executive sponsor, program governance board, portfolio management team, program's beneficiaries, and other stakeholders

within the program, and the benefits from the projects and those of the overall program are sustained (OGC 2007).

Table 11.2 summarizes different attributes of stage gate reviews.

PERIODIC HEALTH CHECKS

Because of certain contextual factors and program attributes, such as the extended life cycle of a program, there might be a need to conduct reviews between the stage gates (Müller 2011). Such reviews can be planned in advance and should be aligned with the information needs of the governance board and governance needs of the program.

PMI (2013) considers periodic health checks to be an important governance mechanism especially when there is an extended time period between scheduled phase gate reviews. The focus of these reviews is to monitor the program's performance and benefits delivery.

Table 11.3 summarizes different attributes of periodic health checks.

There are cases where a program review request can come from the program manager to resolve issues and respond to risks. These reviews can be considered as on-demand consulting and review sessions and are discussed in the next section.

TABLE 11.3

Periodic Health Checks—Attributes

Session type	Preplanned sessions based on dates, milestones, or events
Session focus	Major focus on program's progress and benefits delivery
Target participants	Program management team, executive sponsor, program governance board, and program's beneficiaries

TABLE 11.4

On-Demand Consultation and Review—Attributes

Session type	On-demand
Session focus	General view of different aspects of the program such as progress, risks, issues, decisions, and other desired areas
Target participants	Program management team, executive sponsor, and governance board members

ON-DEMAND CONSULTATION AND REVIEWS

The program might require immediate and direct input from the governance team in terms of direction, guidelines, and feedback related to certain aspects of the program activities and decisions, for example, an opportunity to buy certain goods at discounted rates before a deadline or in bulk. In addition, based on changing organizational context, the governance team might want to provide their immediate input to realign the direction of the program. The program governance plan specifies the requirements and criteria for scheduling such sessions.

To ensure that the program responds to the evolving context, such as the one mentioned above, in an efficient and effective manner, the program governance team permits on-demand consultative sessions. These sessions are not planned in advance; however, the mechanism is put in place to accommodate them as and when required.

Even though such decisions can be made during the stage gate or strategy review sessions, in some cases the program team requires immediate input and action. This is to ensure that time-bound opportunities are exploited through timely decisions.

Table 11.4 summarizes different attributes of on-demand sessions.

PROGRAM AUDITS

Program audits are governance mechanisms that are focused toward ensuring program compliance with processes and standards. The objective of a program audit is to warrant that the program is being managed according to the standard processes, and the program finances are being spent in compliance with the plan (PMI 2013).

Audits are conducted by internal audit departments or by external entities, such as governmental regulatory authorities. The scope and objectives may vary depending on the type of audit being conducted. Every audit activity is planned, and the audit plan ensures that:

1. The audit objectives and scope are defined.
2. The auditor is aware of the program environment.
3. Standardized audit processes are documented.
4. The auditor has a clear understanding of the program processes and procedures that have to be reviewed.
5. In case of financial audits, the auditor recognizes the financial processes used and has identified benchmarks for comparison.
6. The resources required for the audit have been identified and recruited.
7. The roles and responsibilities of the resources are defined and communicated.
8. The deliverables from the audit process have been identified.
9. The audit communication plan has been formed and communicated.
10. In case of planned or announced audits, whether internal or external, the schedule of the audits is developed and communicated to the program team.
11. The procedure for documenting and communicating the audit results are specified.

Program audits consume time and resources. To ensure that the program resources are not overloaded with the additional responsibility of preparation for audits, the audit committee provides assistance and guidance to the program team in the form of mentoring, training, and on-job assistance. It is the responsibility of the program manager to be prepared for audits, whether they are preplanned, announced, or sudden. However, it is a governance responsibility to oversee and ensure this preparedness.

The objective of audits is to make sure that there are no unauthorized deviations from the plan or authorized procedures, that is, all deviations are approved at the required authority level. Any unapproved deviation from the plan is reported to the governance entities and the executive management. Internal program audits keep the program aligned and compliant with the organization's standards, which can help organizations to respond to any external compliance and audit needs. The implementation

TABLE 11.5

Program Audits—Attributes

Session type	Planned, announced, or on-demand sessions
Session focus	Specific focus on compliance with defined processes and standards and the program's financial procedures
Target participants	Management team, executive sponsor, governance board members, program management office members, audit committee, internal audit department, and external authorities

of an audit function ensures that the program delivers quality results and helps avoid the need for later corrective actions.

Table 11.5 summarizes different attributes of program audits.

PLANNED AND ON-DEMAND REPORTING

As part of the governance framework the program governance board receives reports and communications from the management team and, in turn, provides reports to the executive management and other relevant stakeholders.

The schedule and mode of communication from the management team depends on the communication management plan and is a management level responsibility. Based on the program's governance plan, the received reports are reviewed, analyzed by the program governance board, and executive reports are generated. These reports are disseminated to the program stakeholders based on the defined schedule.

There is, however, always a need for on-demand reporting, as different program stakeholders, such as external regulatory authorities, may request additional communication and reports at different stages of the program life cycle. It is the responsibility of the program's governance team to ensure that such reports are created and submitted to relevant entities.

Most of the reporting is done based on the program communication management plan or governance plan; thus, this can be considered as a planned activity. In case the program's governance team receives continuous requests for additional reporting, it is imperative to review the communication plan and revise it accordingly. The participants of this

TABLE 11.6

Planned and On-Demand Reporting—Attributes

Session type	Mostly planned, however, can be on-demand as well
Session focus	General view of different aspects of the program such as progress, risks, issues, decisions, and other desired areas
Target participants	Management team, executive sponsor, governance board members, program management office members, and sometimes external authorities

reporting mechanism include the management team, executive sponsor, governance board members, program management office members, and, sometimes, external authorities.

Table 11.6 summarizes different attributes of program audits.

QUALITY REVIEW

Quality review sessions are focused on ensuring that the program is being managed, and the benefits are being delivered, according to the defined quality standards. These standards are specified in the quality management framework designed by the program governance board.

Any deviation from the specified standard is recorded, and corrective actions are recommended. These sessions also assist in identifying preventive actions that assist in ensuring that the program does not deviate from the defined quality measures. The program's governance board typically schedules these sessions at regular intervals so that the aspect of quality is built into deliverables, and benefits are achieved.

These sessions can be conducted as:

1. Regularly scheduled sessions, which have a defined scope and agenda that is aligned to the program's quality framework. This will typically include a review of the implementation of quality standards and a review of the deliverables or benefits from a quality perspective. The schedule and agenda can be reviewed and updated based on the changing program context.
2. Based on certain events, such as a request for a component transition or a need for integration of component deliverables, where the focus will be more on the quality of the deliverable rather than the process of achieving the outcome.

TABLE 11.7

Quality Review—Attributes

Session type	Preplanned sessions based on dates, milestones, or events
Session focus	Specific focus on program's quality aspects
Target participants	Management team, executive sponsor, governance board members, beneficiaries, advisory panel members, and audit committee

During a quality review session, the program team will receive instant feedback from the program governance board and relevant stakeholders. Table 11.7 summarizes different attributes of quality review sessions.

ADVISORY SESSIONS

Advisory sessions are conducted to allow the program management team to engage with the advisory panel to discuss specific issues or for general guidance related to the program activities. The program's governance board typically schedules these sessions at regular intervals so that the advisors can review the program's progress from different perspectives and can provide guidance in order to keep the program aligned with the goals.

These sessions can be conducted as:

1. Regularly scheduled sessions, which have a defined scope and agenda. The scope and agenda can be decided by the program's management and the governance team based on requirements. The schedule and agenda can be reviewed and updated based on the changing program context.
2. On-demand sessions based on requirements from the program team, program's governance team, or the advisory panel to discuss certain inquires or scenarios or to just discuss certain solutions and ideas.
3. Ad hoc advisory call to discuss some specific issues that need expert advice.

During an advisory session, the program team will receive instant feedback from the experts. These sessions are designed to tackle specific questions, challenges, or strategies.

Table 11.8 summarizes different attributes of advisory sessions.

TABLE 11.8

Advisory Sessions—Attributes

Session type	Preplanned sessions based on dates, milestones, or events. Can sometimes be on-demand based on specific needs of the program team
Session focus	General view of different aspects of the program such as progress, risks, issues, decisions, and other areas, where advice is needed by the program team
Target participants	Program management team, executive sponsor, governance board members, and advisory panel

SUMMARY

Program governance exists within the corporate governance framework. The governance functions, defined earlier, are executed by the governance entities through implementation of a governance mechanism. The governance mechanism for a particular program depends on multiple factors. However, one thing that needs to be ensured is implementing the right balance between delegation of responsibilities to management and governance.

There should be some mechanisms, such as stage gate reviews or periodic health checks, that have to be executed periodically or at important milestones of the program. Mechanisms such as strategic reviews can be conducted in alignment with the organization's strategic review session. In some large-scale programs, certain entities should be formed, such as audit and advisory committees, which should provide guidance and direction to the program team at different stages during the program life cycle.

The program's governance team should not get involved in the day-to-day management of the program. Rather, the governance team should oversee the program activities to ensure that the program delivers its objectives and resulting benefits.

The diagram in Figure 11.1 summarizes the program governance mechanism grouped according to themes discussed earlier.

Stretton (2010) proposed different aspects of governance which should be kept in mind while designing a governance mechanism. The organization should:

1. Establish a governance framework, which should include defining the structure and membership of program governors, policies and

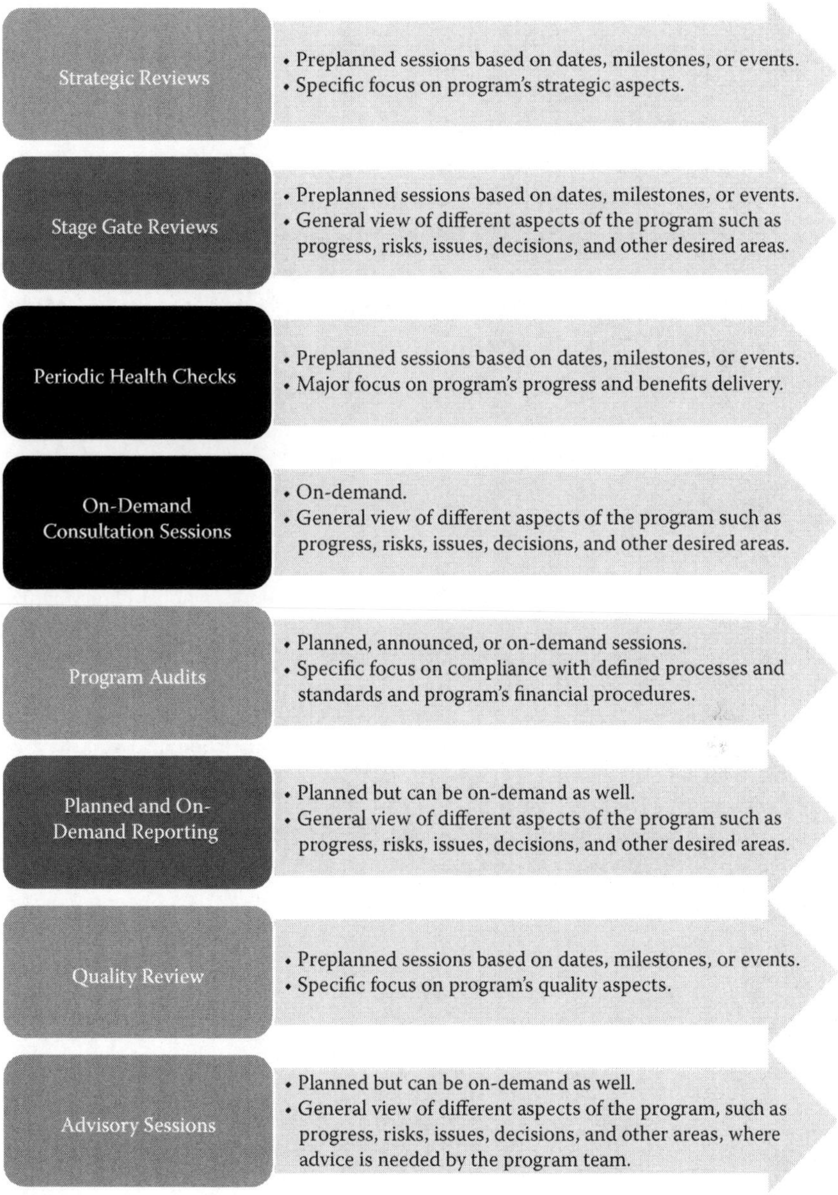

Strategic Reviews
- Preplanned sessions based on dates, milestones, or events.
- Specific focus on program's strategic aspects.

Stage Gate Reviews
- Preplanned sessions based on dates, milestones, or events.
- General view of different aspects of the program such as progress, risks, issues, decisions, and other desired areas.

Periodic Health Checks
- Preplanned sessions based on dates, milestones, or events.
- Major focus on program's progress and benefits delivery.

On-Demand Consultation Sessions
- On-demand.
- General view of different aspects of the program such as progress, risks, issues, decisions, and other desired areas.

Program Audits
- Planned, announced, or on-demand sessions.
- Specific focus on compliance with defined processes and standards and program's financial procedures.

Planned and On-Demand Reporting
- Planned but can be on-demand as well.
- General view of different aspects of the program such as progress, risks, issues, decisions, and other desired areas.

Quality Review
- Preplanned sessions based on dates, milestones, or events.
- Specific focus on program's quality aspects.

Advisory Sessions
- Planned but can be on-demand as well.
- General view of different aspects of the program, such as progress, risks, issues, decisions, and other areas, where advice is needed by the program team.

FIGURE 11.1
Program governance mechanisms.

procedures for governance, defining authority of different roles, and the decision-making process.

2. Set up criteria for acceptance and assessing the projects for acceptance.
3. Ensure resource availability for projects and programs, which is related to the support dimension.
4. Review project progress in terms of cost, timeline, and benefits delivery at a predefined point or otherwise.
5. Resolve escalated issues that are not manageable at the level of projects at the program level or by the program board.

Factors influencing the governance framework design are discussed in the next chapter. This is where the importance of contingency of governance framework comes into play.

REFERENCES

Hanford, M. F. 2005. *Defining Program Governance and Structure*. IBM. Retrieved on March 13, 2013 from http://www.ibm.com/developerworks/rational/library/apr05/hanford/.

Müller, R. (2011). Project governance. In Pinto, J., Morris, P., and Söderlund, J. (Ed.), *Oxford Handbook of Project Management*. Oxford, UK: Oxford University Press.

Office of Government Commerce (OGC). (2007). The OGC Gateway™ process: A manager's checklist. *OGC Best Practice—Gateway to Success*. London, UK: Office of Government Commerce.

Project Management Institute (PMI). (2013). *The Standard for Program Management* (Third Edition). Newtown Square, PA: Project Management Institute.

Stretton, A. (2010). Note on program/project governance. [Featured Paper]. *PM World Today*, *XII*(I), 1–14.

12

Factors Affecting Program Governance

"I need to get approval for this," replied the program director. In 2010, I was part of an advisory panel where our main goal was to retrack a competence enhancement program (CEP) for a certain government entity in the Gulf Cooperation Countries (GCC). This program included multiple initiatives, each of which was focused toward improving the efficiency of the department in different areas such as:

1. Organizational structuring
2. Software development
3. Tactical operations

These initiatives were aligned together under a CEP, which had a vision to improve the department's operational efficiency, resulting in better services to its customers. One of our recommendations was to improve the visibility of the program by placing banners in the department that had program objectives inscribed on it. The cost of implementing this suggestion was approximately $2,700, which was less than 0.0015% of the program's actual budget. But the program director had to approve this amount even though he had this buffer available. Upon inquiry he mentioned that according to the framework designed in 1998, the program directors had a spending power of $1,000 without approval of the board! No one had attempted to revisit the framework for twelve years, resulting in an approval authority misaligned with programs in the current context. Needless to say, getting the approval took the program director four weeks. If the governance framework's design team considered different factors such as prevailing market conditions while designing a governance framework for CEP, this valuable time could have been saved.

Corporate governance has a significant impact on the governance frameworks of the contained temporary organizations. However, there are various other factors which are unique to programs and have an impact on program governance framework. This chapter discusses different factors that influence the program governance framework.

This influence can be on any element of the program governance framework, whether identification of program governance domains, which governance functions to apply, or what should be the governance structure or the design of the governance mechanism.

ORGANIZATIONAL FACTORS

The governance of program comes under the umbrella of corporate governance, which is governed by the industry under which the organization is operating with each industry having its own set of governance requirements. Thus, the same governance regime cannot be applied uniformly to all types of programs as different programs might require different regulatory compliance. Winch (2001) states that "the range of governance options open to any firm is limited by the institutional context within which it trades" (p. 799).

Corporate Governance

Crawford et al. (2008) look at corporate governance as more of a stable function in organizations, whereas the governance of a project/program may differ from one another. However, there should be some sort of consistency between them in the sense that the governance of a project/program should not be in conflict with the governance needs of the organization. They look at compliance and audit aspects of corporate governance and treat support as a separate function (Crawford et al. 2008). Both of these functions in combination form the basis on which the responsibilities of project/program governance, in general, and the sponsor, in particular, are defined.

While focusing on projects, Müller (2011) mentions that agreements and contracts are made ex ante project planning and execution, and act as a basis for governance setup for the project. These contracts should abide by the corporate governance function of the organization. In case the project is being managed, governed, or executed by multiple entities, then the governance function of the project (as defined in the contract or agreement) must abide by or conform to the corporate governance structure of the participating parties (Müller 2011). This ensures that the governance requirements for each entity are taken care of while setting up the governance regime of the project. This also defines the behavior or conduct expected from individuals coming from separate entities when they become part of the project.

It is clear that the governance of a program can differ based on various factors such as the type of the program or the number of participating entities; however, the governance function should be aligned with the corporate governance policy of the participating parties.

The influence of corporate governance is the strongest during the formation and organization stages; however, this impact remains throughout the program life cycle. This factor influences all aspects of program governance, that is, program governance scope, governance functions, governance structure, as well as the governance mechanisms.

Organizational Governance Paradigm

The organizational orientation and focus has a major role to play in designing corporate governance as well as the program governance framework. Weil and Ross, while mentioning corporate and IT governance frameworks, refer to this as the desirable behavior in the organization (Weil and Ross 2004).

In relation to the organizational orientation, Turner and Keegan (2001) mentioned that hybrid governance structures are applied by project-based organizations for the governance of temporary initiatives. However, when one looks at the details of the governance structure, the number of people performing the governance role of the broker and steward as well as their responsibilities differs based on differences in the project settings. This difference is because organizations tend to structure the governance mechanism based on attributes of these initiatives so that economies of scale can be achieved (Turner and Keegan 2001).

Müller has carried out research in this domain, while focusing on the project governance regimes. Müller (2010a) mentions that the organizational governance paradigm acts as a major influencing factor in designing the project governance as well as the project management governance framework. The organizational governance paradigm is

TABLE 12.1

Organizational Governance Paradigm

	Increasing Value for the Shareholder	Increasing Value for Stakeholder
Outcome-based mind-set	Flexible economist	Versatile artist
Focus on compliance	Conformist	Agile pragmatist

Source: Based on Müller R. (2010b). Project governance. [Monthly Column—Series on Advances in Project Management]. *PM World Today, XII*(III), 1–6.

based on whether the organization focuses on stakeholders or shareholders. It is also dependent upon whether the organization pays more attention to controlling outcomes/results or to compliance (Müller 2010a,b). Based on these two attributes, Table 12.1 mentions different governance paradigms that an organization can have (Müller 2010a; Müller 2011).

Müller (2010b) mentions that conformist and agile pragmatist governance paradigms focus on behavioral control through compliance with processes and practices. The conformist paradigm focuses on increasing the shareholder returns and is an ideal approach when the processes and outcomes are well defined and linked together. The agile pragmatist works toward increasing the collective benefits for all the stakeholders over a period of time. Organizations with this approach have to adapt to the changing needs of the stakeholders (Müller 2011).

The flexible economist and versatile artist paradigms are more concerned with the outcomes and require project management as a core competency. However, the versatile artist has to work toward managing the expectations of a wide range of stakeholders through balancing diverse requirements and setting up tailored processes; whereas the flexible economist has to adapt to increase the value of outcomes for shareholders, such as a return on investment (ROI) (Müller 2010a).

As an example, an organization with strict attention to conformance with processes cannot have a governance framework, which allows changes to the program baselines without completion of the required prerequisites.

The influence of the organizational governance paradigm remains throughout the program life cycle. This organizational governance paradigm influences all aspects of program governance, that is, program governance scope, governance functions, and governance structure, as well as the governance mechanism.

EXTERNAL FACTORS

Factors outside the boundaries of the program have a strong influence on the program's governance framework. PMI (2013) lists certain environmental factors, such as business environment, market trends, legislation, industry, economy, cultural diversity, and others, which should be considered when performing ongoing program assessment and alignment functions.

Cultural Influences

The internal culture of the organization as well as external culture of the society plays a significant role in the design of a program governance framework. While investigating three public investment projects in the United Kingdom and Norway, Klakegg et al. (2008) identified the differences in the approach, structure, and embedded governance principles as well as the governance mechanism based on the differences in society and culture. As an example, civil servants in Norway are not accountable for their actions, and the government owns responsibility; thus, the Norwegian governance framework focuses more on control. Whereas in the United Kingdom, civil servants, such as senior responsible owners (SROs), are held accountable; thus, the focus is more on guidelines and internal assessments (Klakegg et al. 2008).

Cultural aspects also influence all aspects of program governance, that is, program governance scope, governance functions, and governance structure as well as the governance mechanism.

Legislation, Regulations, and Standards

Organizations operate within the regulation and legal boundaries defined by legislative and regulatory authorities set up by the government to ensure that organizations operate under a framework of control mechanisms, accountability, decision-making process, and clear distribution of power. It also ensures that organizations refrain from getting involved in unethical practices that can negatively impact the shareholders, in particular, and stakeholders, in general. Any change in legislation can strongly influence the direction of the program or in some cases the organizational strategy.

In addition, there are standardization authorities that develop policies and standards particular to different industries. The International Organization for Standardization (ISO) is one such entity that is focused on developing international standards that provide specifications, policies, and guidelines for products, services, and best practices. Based on the industry in which the program operates, it may have to comply with certain standards and practices.

Some of these influences impact the organization directly, and thus are generally handled through the overarching, corporate governance framework. However, others may have a direct impact on the program. An example would be a program that has an objective to develop a new bridge, in a densely populated location, which will result in displacement of the inhabitants to a new area. The aspect of ensuring that the impacted stakeholders are aligned with the greater benefit that the program is delivering, the program governance entities should ensure frequent communication and stakeholder buy-in through different activities.

Legislation, regulations, and conformance to standards influence all aspects of program governance, that is, program governance scope, governance functions, and governance structure, as well as the governance mechanism.

PROGRAM ATTRIBUTES

Certain attributes, based on which programs can also be classified, have an impact on the program governance mechanism. This impact has to be identified and understood to ensure that the program's governance framework is in alignment with the program's governance need.

Benefits Delivery Mechanisms

The mechanism of a program's benefits delivery will have an impact on the program governance framework. The program governance framework has to align with the benefits delivery model. An example would be programs that are delivering benefits in an iterative model might require a reduced number of periodic health checks as compared to programs delivering benefits at the end of their life cycle. This is because these programs will already have a larger number of stage gate reviews that will be conducted at the component transition stages. The program's governance mechanism should support consistent transition of incremental benefits to recipients and should also influence the recipient's readiness to sustain the delivered benefits. The focus of governance in such a delivery model will be on integration and frequent quality reviews.

A program delivering benefits at the end of its life cycle might require an entirely different governance framework, such as a higher number of periodic health checks and review sessions.

This factor influences program governance functions, institutions, and roles as well as the program's governance mechanism. This influence generally occurs at the formation and organization stages of the program life cycle, that is, when the program benefits are being defined and the program benefits delivery mechanism is being determined.

Program Structure

The number of entities involved in the delivery of the program has an impact on the manner in which the program governance structure is defined. Also, the structure of the program in terms of the delivery team and beneficiaries, that is, whether the program is delivering benefits internally or to an external customer, influences the roles and responsibilities, as well as the communication needs of the program stakeholders.

An example would be multi-owned programs. Program governance concerns in such arrangements are focused around:

1. Alignment of interest and understanding between the parties.
2. Owner nominations.
3. Shared setup for program management.
4. Disclosure and reporting arrangements aligned with requirements of all parties.
5. Risk and reward mechanisms for the program life span.
6. Mechanisms for joining and leaving the multi-owned program setup.

As discussed, another aspect would be whether the delivery organization is internal or external to the beneficiary organization. The type of contract, level of trust, and the experience of the vendor, in addition to other factors, will have an influence on the governance framework. In the case of external vendors, from a principal–agent perspective, problems of moral hazards and adverse selection have to be resolved through developing formal contracts, which address these issues. Programs in which the beneficiary and delivery team belong to the same organization will have a different approach toward governance than the programs that have the beneficiary and delivery teams belonging to a different organization.

Differences in organizational cultures and corporate governance requirements of participating entities will also have an influence on the program governance framework. This factor has a strong influence on the governance institutions and roles, and also influences the program's governance functions and mechanisms. This influence generally occurs at the formation and organization stages of the program life cycle, as that is the time when the contractual aspects of the program are handled.

Program Uncertainty and Complexity

Uncertainty and complexity associated with the program act as a major factor in determining the governance structure (Turner and Keegan 2001). Miller and Hobbs (2005) mentioned that large complex projects have a very high level of uncertainty in the beginning, and thus they require different governance regimes.

Program complexity can be assessed from the following perspectives:

1. Technical complexity, which specifies the level of technical difficulties and challenges that the program will face in delivering the benefits. This aspect is also dependent on the novelty of program deliverables and benefits.
2. Scope complexity, which specifies the number of components involved and the integration between them to deliver the benefits.
3. Stakeholder complexity, which specifies the number of stakeholders involved and their influence on the program.
4. Impact complexity, which specifies the type and level of changes and impact that the program will have on the beneficiary organization.

Project uncertainty can be assessed from the following perspectives:

1. Requirement certainty, which defines how clear the program scope is and how well defined the program objectives and expected benefits are.
2. Requirement stability, which defines whether the program requirements are relatively stable or are consistently evolving.

3. Process certainty, which defines how clear the program's processes are and the program road map that will lead to the delivery of program benefits.

High uncertainty and complexity of the program's outcome, process, and road map may influence the program governance team to intensively monitor and control the program's progress and activities. As the program's vision becomes clear, that is, the overall uncertainty and complexity of the program lessens, the program governance team may tend to rely more on the management team and will perform the oversight function with a less probing attitude.

These factors have a strong influence on the program governance team's attitude toward different scope elements. Additionally, this also influences interrelated governance aspects of functions, institutions, and governance mechanisms. The influence of this factor remains throughout the program life cycle.

Program Life Cycle Stage

The program governance framework should be adaptable to the program's life cycle as different phases of a project require a different focus from the governance regime. As the program unfolds, the types of issues and challenges also changes based on the program context, requiring a flexible and adaptive governance framework. This is because the surrounding environment of the program is bound to change, requiring a review and update to the program governance framework.

Programs have a high level of uncertainty and complexity, which is resolved through progressive elaboration. During the formation phase of the program the focus is on benefits identification and program scope definition. Through the organizing stage the focus shifts to defining the program structure, plans, and laying out the mechanism of management and governance. When the program moves to the benefits delivery stage the governance entities focus more on monitoring, supporting, and controlling the different elements of program scope. Finally, the closure phase of the program involves assessing the quality of the program deliverables and the success of the program is determined.

Different stages of the program life cycle may require a different level of emphasis on program governance scope elements. In addition, the structure of the program governance model may be revised as the roles and responsibilities of different governance entities may be revised. Finally, the program's governance mechanism may be updated as well based on changing needs of the program; for example, programs might not require frequent audits during the formation and organization stages whereas the need for audits might increase during the benefits delivery and closure phases.

Program Strategic Value

This factor determines the strategic value of the program under consideration. The strategic value of the program is determined from the following perspectives:

1. The strategic importance of the project.
2. The investment the organization is making in the project.

These combine to provide organizations with strategic value perspective, which can assist in determining the strategic priority of the program.

A program's strategic value can be considered as an influential attribute. Strategic decisions are made by senior managers, where programs act as strategy execution vehicles. Because of the investments made by the organization in highly strategic projects and programs, the governors focus on these initiatives to ensure that the programs meet their desired objectives. The information flow in the case of strategic value is from the senior managers to the program manager and the program team. Program governors have to ensure that the program stays aligned with the organizational goals, and the investment made in these programs delivers the required value to the organization.

As the strategic value of the program decreases, that is, if the strategic importance of the program is low or the cost of the program is low, the focus on different dimensions of governance also decreases. This may be because the investment required to govern such programs may be much higher than the program's value to the organization. This is in alignment with the classical transaction cost economics perspective, which advocates economization of governance structure based on the transaction attributes (Williamson 1979). This means that it does not make economic sense to create governance frameworks with complex structures, multifacet measurement parameters, and tools for programs of low strategic value. Rather, a simpler model can be created that can ensure program delivery without creating cumbersome bureaucracy.

High strategic value programs require a higher level of governance from the governors to ensure that the program can deliver its strategic value. Even though the program governors allow the project manager to manage and execute the program activities, they work closely with the program managers and the program team. They advocate for the program in front of other stakeholders and the program team whenever needed. They also create monitoring mechanisms in such a way that any deviation above the threshold is observed, and decisions are made to bring the program back on track in a timely fashion. All this is done to ensure that the program delivers its objectives, which are critical for organizational success.

Program Performance

While referring to the sponsor role in project/program governance, Crawford et al. (2008) mention that the focus of the sponsor should shift from support dimension to scrutiny and control in the case where the project/program is facing issues or is performing poorly. On the other hand, if the project/program is moving smoothly or exceeding expectations the focus generally moves toward support and guidance (Crawford et al. 2008).

It is important to identify program performance trends and control program performance related issues by escalating them to executive management for actions, especially when they cannot be handled at the management level. This factor has a strong influence on the program governance entity's focus on different functions of governance. The performance of the program may also influence the mechanism of governance, additional health checks if the program is performing poorly, and the roles and responsibilities of the governance entities. The influence of program performance starts with the organization stage. However, the impact of this factor is strongest during benefits delivery and closure stages, that is, where the program is delivering its objectives and resulting benefits.

DIFFERENCES IN GOVERNANCE REGIMES—SOME RESEARCH

This book proposes certain concepts that can be applied in most organizations, most of the time, while implementing a program governance framework. However, the actual implementation of the governance practices for a specific program may differ based on program types and other contextual factors, such as the ones discussed above. This concept is also in alignment with the idea proposed by Shenhar (2001), when he mentioned a similar concept at the project level and stated that all types of projects cannot be managed using a standard, universal project management mechanism.

Turner and Keegan (2001), while reviewing governance mechanisms, mention that project-based organizations require new approaches to manage transient, novel, and unique work that they undertake. This is because a project's bespoke product, which has uncertainty built at its core, requires novel processes and approaches as the entire value chain of producing the product is uncertain, interlinked, and dependent. They focused on the structural perspective of governance and mention that a hybrid governance structure is adopted for all projects in a client–vendor relationship. However, the roles and responsibilities of governance actors will differ between organizations based on the type of undertaking. This difference is attributed to the organizational objective to reduce transaction costs in order to achieve economy of scale.

Müller and Blomquist (2006) described the impact of project attributes and the organizational complexity on the roles and responsibilities of middle managers while governing projects through portfolios and programs. While referring to program and portfolio management practices, they mention that these two approaches are different yet parallel mechanisms required for governance of project management in the organization. Program management focuses on the interdependencies between projects in terms of shared objectives, whereas portfolio management relates to the interrelationship between projects because of shared management needs (Müller and Blomquist 2006).

Their research was comprised of a qualitative study and a subsequent confirmatory quantitative study, and the results were later triangulated (Müller and Blomquist 2006). Program and portfolio management practices, roles, and responsibilities (dependent variables) were tested against the type of projects and environmental complexity (independent variables). Environmental complexity included the complexity of the decision-making process as well as the stability of the environment for making decisions. Project types were judged based on whether the project delivery was internal to the organization or external, the duration of the project, and whether the end result was a product or a service.

The results of the study showed that:

1. The practices for program and portfolio management are dependent on the environmental complexity as well as the type of project. Higher complexity requires use of specific program and project management processes and tools. At the same time, it requires clarity in the roles of middle managers for different governance responsibilities. Projects that required organizational change required more effort in resource planning, project prioritization, and issue management. Projects of longer duration required more planning effort in terms of business planning as well as planning reviews and poor project identification.

2. The dynamic nature of the organizational complexity did not impact the program and portfolio management mechanisms. Also, the project categorization based on whether the end result of the project is a product or a service had no impact on governance practices, roles, and responsibilities.

3. There was also a difference in the mechanisms (roles, responsibilities, and practices) of program and portfolio management between high- and low-performing organizations. High-performing organizations seem to have more vigorous practices of program and portfolio management as compared to their low-performing counterparts. High-performing organizations also have practices related to poor project selection. Also, project managers in high-performing organizations have to report to higher authorities, whereas in low-performing organizations, they generally report to middle managers.

SUMMARY

While designing and implementing program governance mechanisms various factors have to be considered. Corporate governance has a major influence; however, the type of program and its performance strongly influence the manner in which governance is implemented in programs.

All programs cannot be governed using the same mechanism because of the highly uncertain nature of programs and their constituent projects. The governance regimes of programs can be seen as self-organizing and dynamic, which means that they have to adapt and evolve according to

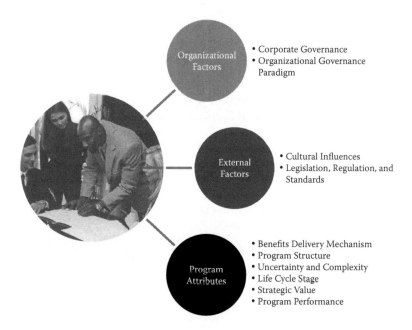

Organizational
Factors

• Corporate Governance
• Organizational Governance
  Paradigm

External
Factors

• Cultural Influences
• Legislation, Regulation, and
  Standards

Program
Attributes

• Benefits Delivery Mechanism
• Program Structure
• Uncertainty and Complexity
• Life Cycle Stage
• Strategic Value
• Program Performance

FIGURE 12.1
Program governance—influential factors.

the changing program context. This will ensure an effective governance framework instead of mindless bureaucracy following needless processes.

There is alignment between the concepts proposed in this section and the classical theory of transaction cost economics. If we equate programs and their constituent projects as transactions, we can align with the governance mechanism proposed by Williamson (1985), which states that "Transaction costs are economized by assigning transactions (which differ in their attributes) to governance structures (the adaptive capacities and associated costs of which differ) in a discriminating way" (p. 18).

The diagram in Figure 12.1 summarizes the factors discussed in this chapter that have to be considered while designing efficient and effective governance frameworks for programs.

REFERENCES

Crawford, L., Cooke-Davies, T., Hobbs, B., Labuschagne, L., Remington, K., and Chen, P. (2008). Governance and support in the sponsoring of projects and programs. *Project Management Journal, 39*(1), 43–55. doi: 10.1002/pmj.20059.

Klakegg, O. J., Williams, T., Magnussen, O. M., and Glasspool, H. (2008). Governance frameworks for public project development and estimation. *Project Management Journal, 39*(1), 27–42. doi: 10.1002/pmj.20058.

Miller, R. and Hobbs, B. (2005). Governance regimes for large complex projects. *Project Management Journal, 36*(3), 42–50.

Müller, R. (2010a). Project governance: Procedural straight jacket or freedom of arts. [Presentation at the CONCEPT Symposium, Oscarsborg, Norway, September 16–17].

Müller, R. (2010b). Project governance. [Monthly Column—Series on Advances in Project Management]. *PM World Today, XII*(III), 1–6.

Müller, R. (2011). Project governance. In Pinto, J., Morris, P., and Söderlund, J. (Ed.), *Oxford Handbook of Project Management*. Oxford, UK: Oxford University Press.

Müller, R. and Blomquist, T. (2006). Practices, roles, and responsibilities of middle managers in program and portfolio management. *Project Management Journal, 37*(1), 52–66.

Project Management Institute (PMI). (2013). *The Standard for Program Management* (Third Edition). Newtown Square, PA.

Shenhar, A. J. (2001). One size does not fit all projects: Exploring classical contingency domains. *Management Science, 47*(3), 394–414. doi: 10.1287/mnsc.47.3.394.9772.

Turner, J. R. and Keegan, A. (2001). Mechanisms of governance in the project-based organization: Roles of the broker and steward. *European Management Journal, 19*(3), 254–267.

Weil, P. and Ross, J. W. (2004). *IT Governance: How Top Performers Manage IT Decision Rights for Superior Results*. Boston, MA: Harvard Business School Press.

Williamson, O. E. (1979). Transaction cost economics: The governance of contractual relationships. *Journal of Law and Economics, 22*(2), 233–261.

Williamson, O. E. (1985). *The Economic Institutions of Capitalism*. New York: The Free Press.

Winch, G. M. (2001). Governing the project process: A conceptual framework. *Construction Management and Economics, 19*(8), 799–808.

13

Contingent Governance Framework for Programs (CGFPrg™)

Consider yourself to be an academic researcher who takes his seven-year-old son to a Detroit Tigers game. Your son asks you for lemonade and when you buy it at the concession stand, you are handed a bottle of Mike's Hard Lemonade. You have never heard of it before and you do not know that hard means alcohol, thus you hand it over to your son. Suddenly you see a ballpark security guard running toward you and taking your son away because he had seen him drinking from the bottle. Your son is taken by an ambulance to the Children's Hospital where doctors find no trace of alcohol in his blood. Doctors allow you to take him home but the police officers put your son into a Wayne County Child Protective Services foster home. After three days the court judge rules that your son can come home to his mother, but only if you are moved into a hotel. You are finally reunited with the family after two weeks of humiliation and ordeal.

This real life event happened to Christopher Ratté and his son Leo. The case was published on the front page of the *Detroit Free Press*, was reported on CNN, and was the subject of a Scott Simon commentary on National Public Radio (NPR) *Weekend Edition*. The father was a professor of archeology at the University of Michigan and had never heard of Mike's Hard Lemonade as a drink and had no idea that it contained 5% alcohol. He said that he would have never handed such a drink to his son had he known that it contained alcohol.

At each step the government officials, police officers, social workers, and judges said they hated to do what they did but they had to follow procedures. They wanted to let Leo go home, but the procedures did not allow them to do so. They did not want the father to be separated, but they had to obey the rules.

That is what enforced following does. Please do not get me wrong. I am not against following procedures but doing so without applying the lens of wisdom can do more harm than good. That is what I call *mindless bureaucracy*.

Barry Schwartz, in his TED Talk "Our Loss of Wisdom," filmed in February 2009, mentioned that real-world problems require us to improvise. A wise person knows when and how to make an exception to the rule and how to improvise when required. Barry Schwartz called for a return to practical reasoning or wisdom.

However, improvising and reasoning can only be done when the framework allows you to adapt or to make decisions based on the context. That is where the idea for Contingent Governance Framework for Programs (CGFPrg) is useful. Such frameworks adapt, through revisions, based on changing program attributes and context.

CONTINGENCY THEORY

Contingency theory mentions that organizations can perform at their optimum level when the organizational setup aligns with the organizational context. The structure and process of an organization must fit its context, including the characteristics of the organization's culture, environment, technology, and the size of the task in order to improve organizational performance. Woodward (1965) states that organizations are more successful when their structures and human relationships conform to their technological situations.

Lawrence and Lorsh (1969) mention that the organizational structure, system, and processes should be consistent with external environmental context. Galbraith (1973) states that to achieve organizational effectiveness, the conditions of uncertainty of the task should be taken into consideration.

In summary, the underlying assumption of contingency theory is that there is no one best solution to organize, and any solution to organize is not equally effective under all conditions. Organizational frameworks should adapt to changing context.

ALIGNING AGILITY AND DISCIPLINE: CONTINGENT GOVERNANCE FRAMEWORK FOR PROGRAMS

Contingent Governance Framework for Programs (CGFPrg) proposes that at any given point in time the governance framework for programs, in terms of scope, function, structure, and mechanism, should be aligned with the program attributes and the program context.

This proposition is in alignment with Transaction Cost Economics (TCE) theory, which suggests that in order to economize on transaction costs, the program governance framework should be aligned with the program's attributes and context. This concept also supports contingency theory, which proposes that in order to have an efficient working environment, there should be an alignment between organizational context, which may include the organization's culture, environment, technology, attributes of the tasks, and the management and governing structure.

Pellegrinelli (1997) mentions that the programs exist in an uncertain, competitive, political, and technological environment and have to evolve based on the organizational strategy and business needs. The objective of creating such a framework is based on the understanding that to be effective, the governance frameworks have to adapt to the evolving program context. This will result in provisioning of effective organizational oversight that will assist the program in delivering its objectives instead of creating additional layers of reporting, resulting in delayed decisions and an inefficient working environment.

The framework has two core components; that is, influential factors (IFs) and the governance framework elements (GFEs). Influential factors have an impact on the manner in which governance framework elements are defined and designed in order to have an effective governance framework. During the program's execution, the IFs and their impact should be reconsidered and the governance framework elements should be refined and redesigned based on the changed program attributes and context. This ensures that the program's governance procedures and practices are aligned with the internal program context and external environment within which the program is being executed. An example would be a program governance framework in which the authority of the program manager to purchase necessary equipment or hire crucial services is reassessed based on the yearly inflation rate prevalent in society.

The following sections provide an overview of the IFs and GFEs that have been discussed earlier in this book.

Influential Factors

As in the previous discussion, various factors influence the design of program governance frameworks, and we call them *influential factors* (IFs). The impact of some of the factors is considered while the governance framework is initially designed. These factors include the program attributes, corporate governance framework, and others. However, there might be other factors, such as program performance, which will not impact the initial design, and their impact will be considered during the program execution.

Table 13.1 shows influential factors (IFs) that have an impact on the program governance framework's design and implementation.

Details related to these factors can be found in Chapter 12 of this book.

Governance Framework Elements

Governance framework elements are different dimensions that have to be addressed while designing the framework. There are four governance framework elements:

1. Governance domains, which identify the different areas of a program that will be governed. Program governance functions are applied to

TABLE 13.1

Influential Factors

Influential Factor	Type of Influence
Corporate governance	Organizational factor
Organizational governance paradigm	Organizational factor
Social and cultural impact	External factor
Legislation, regulation, and standards	External factor
Benefits delivery mechanism	Program attribute
Program structure	Program attribute
Program uncertainty and complexity	Program attribute
Program life cycle stage	Program attribute
Strategic value	Program attribute
Program performance	Program attribute

provide an oversight function over assets that have to be overseen in order to meet the program objectives. These assets include, but are not limited to, program structure and processes, program decisions, and program resources. A full discussion is in Chapter 8.

2. Governance functions define different functions performed by the governance entities. These functions are detailed in Chapter 9 and include activities such as defining the program benefits, designing and implementing the governance framework, ensuring strategic alignment, and other related activities.

3. Governance institutions and roles, which identify the people and groups that should be involved in governance, for example, Sponsor, Program Management Office (PMO), and program governance board. The bodies and entities for governance may differ from organization to organization and type of programs that have to be governed. A detailed discussion is in Chapter 10.

4. Governance mechanism, which defines how the governance will take place such as stage gates and program audits. Similar to the management model, the governance mechanism for programs cannot be applied in a similar manner for all types of programs. Thus, organizations adopt different mechanisms to govern their initiatives depending on the influential factors identified earlier. Chapter 11 discusses different governance mechanisms in detail.

These elements should be carefully designed based on the IFs so that an economically viable and efficient governance framework can be designed, which should result in improved program performance.

GUIDELINES TO IMPLEMENT A CONTINGENT GOVERNANCE FRAMEWORK FOR PROGRAMS

Program governors provide governance oversight on different domains of the program by implementing the right balance of support, control, and surveillance through governance functions. These functions are implemented using the governance mechanisms adopted by the organization. In this section, I am defining the mechanism of developing the governance framework. The activities defined in the section will overlap and should not be considered as sequential engagements.

Before program initiation, the governance framework is designed and employed to ensure that an oversight function is in place from the initial stages. The first step in designing the governance framework is the identification of IFs and their supposed impact on the framework's design. Factors such as a corporate governance framework and an organization's governance paradigm will have a direct impact on the governance framework's design. There might be certain reporting requirements defined in the corporate governance framework that will impact the reporting function of a program's governance function. The program's benefits delivery mechanism may have an impact on how the governance mechanism of stage gate reviews is employed.

After identification of IFs and their perceived impact, the governance framework's design team should identify the domains of the program that have to be governed and define the level of governance required for each domain. This will ensure that governance framework takes into consideration all the areas of the program that need to be overseen and directed. Decisions such as how much oversight is required for each domain are made at this stage.

The second aspect that should be considered at this stage is the identification of the governance functions. These functions depend on the governance domains identified, for example, if the governance team decides that the program stakeholders will not be governed by the governance team and will be controlled solely by the program's management team then the scope of governance functions such as reporting and communication, as well as external interfacing and coordination, will be significantly impacted. In addition to the identification of the governance function, the governance design team should also define the scope of each governance function. This activity will provide answers to questions such as how will the program benefits be identified and defined? Or how will the governance team ensure that the program is consistently aligned with organizational strategy?

The governance functions are performed by the governance team, thus the third dimension of the program's governance plan should define the governance entities. The overall governance structure, which includes governance entities, definition of their roles and responsibilities, and the reporting lines, should be developed. This activity will ensure that the program's governance team will recognize clearly their domain of influence and responsibilities. It will also assist the program's management team and other stakeholders in developing effective communication channels as they will have a clear idea about who/how/when of communication lines and needs.

Finally, the governance mechanisms have to be identified and operationalized. Depending on the program's attributes and context, the governance framework's design team will decide which governance mechanisms are aligned with the program's requirements. A program of high strategic value will require frequent stage gate reviews as compared to a program of lower strategic importance. Accordingly, programs of higher complexity will require advisory boards and will employ advisory sessions for consistent guidance. In summary, the design team will design and employ the governance mechanism based on the program needs.

It is important to note that once the governance model is operational, it should not become something that cannot be redesigned or altered for improvement. Because of the program execution, the overall program context will change, and additional IFs will come into play. One of the core IFs during a program's execution is the performance of the program. A high-performing program will require a different governance mechanism than a program that is performing poorly. Program governors of a low-performing program will require additional reporting and might employ additional gate reviews in order to bring the program back on track. Programs with consistent low performance might qualify for eventual termination.

Other IFs will play a role in redesigning the governance framework so that the objectives of economic viability and efficient governance are consistently achieved. The redesigning of the governance framework may take place at predefined milestones or at the time of a gate review or program audits. This review could also occur during an overall review for greater process effectiveness or during a maturity assessment.

One important aspect to consider is that the governance design team should provide the right balance of agility and discipline. The governance framework should allow program governors and managers a mandate to act based on their discretion within certain thresholds. These threshold limits should be determined based on the IFs and should be revised based on the evolving program context. The governance and management team should be allowed to make certain decisions, within the defined threshold, without getting prior approval from the executive management.

SUMMARY

The role of governance is to oversee the program and ensure that the program either delivers its objectives, or its existence is reassessed. The redesigning of the program governance framework, based on IFs, ensures that governance consistently plays its role effectively and efficiently.

Program governance exists within the corporate governance framework. However, various other aspects, such as influential factors, institutional setup, and governance dimensions must be considered while designing a program governance framework. It is important to consider these factors as they may impact the governance framework design and its implementation. The diagram in Figure 13.1 provides the conceptual view of the Contingent Governance Framework for Programs (CGFPrg).

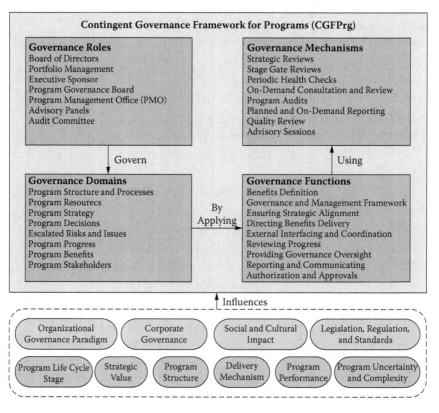

FIGURE 13.1

Contingent Governance Framework for Programs (CGFPrg).

It is once again important to note that program factors such as a program's performance come into consideration during the program's execution and delivery. Thus, it is important to revise the framework based on varying program performance and evolving program context.

The governance framework revised on a regular basis ensures that the program receives the optimum oversight that is required to deliver the program's intended benefits.

REFERENCES

Galbraith, J. (1973). *Designing Complex Organizations*. Reading, MA and UK: Addison-Wesley.

Lawrence, P. and Lorsch, J. (1969). *Organization and Environment*. Homewood, IL: Richard D. Irwin.

Pellegrinelli, S. (1997). Programme management: Organising project-based change. *International Journal of Project Management*, 15(3), pp. 141–149.

Schwartz, B. (2009). *Our Loss of Wisdom*. Retrieved May 12, 2012, from http://www.ted.com/talks/barry_schwartz_on_our_loss_of_wisdom.

Woodward, J. (1965). *Industrial Organization: Theory and Practice*. London, UK: Oxford University Press.

Appendix: Glossary of Acronyms

APM: Association for Project Management
BoD: Board of Directors
CEO: Chief Executive Officer
CoE: Center of Excellence
CPR: Conformance, Performance, and Relating Responsibility
CSFs: Critical Success Factors
CGFPrg: Contingent Governance Framework for Programs
GCC: Gulf Cooperation Countries
GFEs: Governance Framework Elements
IFs: Influential Factors
IP: Intellectual Property
ISACA: Information Systems Audit and Control Association
IT: Information Technology
ITGI: IT Governance Institute
MNC: Multinational Corporation
OECD: Organization for Economic Cooperation and Development
OGC: Office of Government Commerce
PMI: Project Management Institute
PMO: Project Management Office
SOX: Sarbanes–Oxley Act
SPSS: Statistical Package for the Social Sciences
SRO: Senior Responsible Owners

Index

G